Water Resources

Water Resources

Science and Society

GEORGE M. HORNBERGER & DEBRA PERRONE

JOHNS HOPKINS UNIVERSITY PRESS *Baltimore*

Published 2019
Printed in Canada on acid-free paper
9 8 7 6 5 4 3 2 1

Johns Hopkins University Press
2715 North Charles Street
Baltimore, Maryland 21218-4363
www.press.jhu.edu

Library of Congress Cataloging-in-Publication Data

Names: Hornberger, George M., author. | Perrone, Debra, author.
Title: Water resources : science and society / George M. Hornberger and Debra Perrone.
Description: Baltimore : Johns Hopkins University Press, 2019. | Includes bibliographical references and index.
Identifiers: LCCN 2019004930 | ISBN 9781421432953 (paperback : alk. paper) | ISBN 1421432951 (paperback : alk. paper) | ISBN 9781421432960 (electronic) | ISBN 142143296X (electronic)
Subjects: LCSH: Water-supply.
Classification: LCC TD345 .H59 2019 | DDC 333.91—dc23
LC record available at https://lccn.loc.gov/2019004930

A catalog record for this book is available from the British Library.

Johns Hopkins University Press uses environmentally friendly book materials, including recycled text paper that is composed of at least 30 percent post-consumer waste, whenever possible.

CONTENTS

PREFACE

How much water is there on Earth, and how is it distributed over space and time? How much water do humans need, and how much does the environment need? What changes in the future will affect the ability of people to use water sustainably?

Water moves through the global hydrological cycle and supplies the fresh waters contained in rivers, lakes, groundwater, and soil water and that are used by plants and animals. Basic information about the volumes and flows of these water resources is the underpinning for understanding the supply sector for water and is the subject of Part I of this book.

Several primary uses of water that are essential for civilization are to grow food, produce energy, provide needs for drinking, bathing, cooking, and so on, and maintain healthy ecosystems. These needs for water are cornerstones of the demand sector for water. The material for Part II provides a primer on each sector and introduces trade-offs involving reliability, economics, the environment, and water use.

Matching supply and demand is the essence of water resources management. As we move farther into the twenty-first century, pressures on water resources will increase and affect management options. Population growth, climate change, implementations of water law, and deterioration of water quality all will act as constraints. These are topics covered in Part III.

One last question remains: How can knowledge about physical and social systems be harnessed to manage water resources? The final chapter of the book provides an overview of opportunities to manage water sustainably in the future in the face of unavoidable trade-offs.

Our goal is to introduce twenty-first century water resources issues to undergraduate students from across a wide spectrum of disciplines, including natural (environmental) sciences, social sciences, and humanities. This book fits between the two bookends of being highly technical for advanced hydrology, engineering optimization, or law courses and being a completely descriptive summary of issues. It covers basic methods and includes illustrative quantitative calculations and qualitative think-pieces. It covers calculations related to water budgets. It covers data and calculations used to inform decisions about allocation of water resources. It contains case studies that illustrate key principles of water law.

Several conventions are used to make the text user friendly. Each chapter concludes with a summary of key points. Boxes are included in many chapters to illustrate concepts introduced in the main text. Each chapter has a set of example problems based on the material covered. An appendix provides a review of units, dimensions, and conversions useful for addressing each chapter's example problems. And, finally, terms in boldface are included in the glossary.

We are grateful for the comments from our editors, colleagues, anonymous reviewers, and family. We are thankful to our students and teaching assistants for test-driving the material and working through the problem sets.

This book was a true collaboration with each author contributing equally.

I

Water Availability: A Physical Science Primer

CHAPTER 1

The Hydrological Cycle

1.1 Introduction

One of the most stunning things about images of Earth from space is how blue the planet is. Earth truly is the water planet. We see in images not only the large extent of the world's oceans but also ice, snow, and the clouds that deliver water to the terrestrial environment. The presence of water is necessary for life as we know it, and should life be found on other planets, it is almost certain that those planets will be just as blue. The expanse of oceans and the coverage of the land surfaces by river networks is a defining characteristic of Earth (Figure 1.1).

Humans use freshwater resources for many purposes, with the most obvious being personal uses such as drinking, bathing, and sanitation. In addition to these uses, people depend on water resources for many necessary and beneficial purposes, including agriculture, commercial enterprises, mining, recreation, navigation, and thermoelectric and hydroelectric power. From antiquity to the present, water availability and use have been implicit drivers for civilization (Solomon, 2010). Key breakthroughs involving water have directed the course of civilization. These breakthroughs include the rise of large-scale irrigated agriculture some 5,000 years ago, the development of power using waterwheels some 2,000 years ago, and the ability to tame the outbreak of waterborne diseases through proper sanitation a few hundred years ago.

Because water is of key importance to so many essential human activities, it must be managed effectively. Fair allocation and wise use of freshwater resources are significant challenges across the entire world, and conflicts and trade-offs must be negotiated between nations, states, cities, and individual users. Because water is a renewable resource, its availability can appear to be unlimited, particularly in regions that receive plentiful rainfall. Nevertheless,

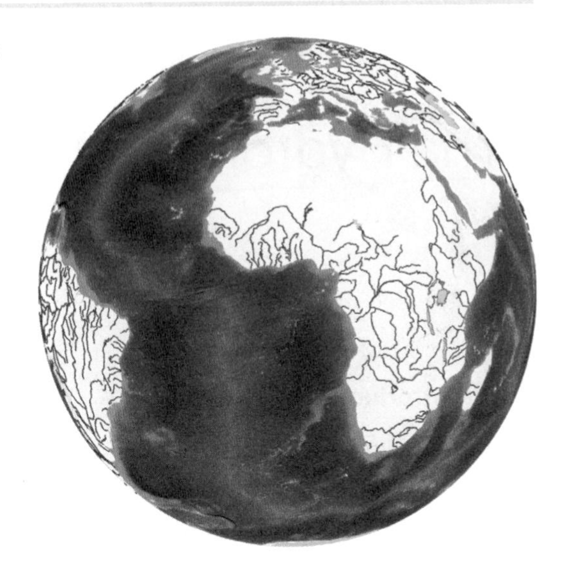

Figure 1.1 A map of Earth showing the oceans and major rivers and lakes of the blue planet.

large growth in the use of water has occurred over the past century, especially over the past several decades. At the same time, many water bodies have undergone increased contamination. To avoid unresolvable water crises in the future, planning for and management of the use of water resources in the twenty-first century is fundamentally important.

A series of simple questions captures some of the basic physical science information we need in learning how to use our water resources wisely. (1) How much fresh water is available on Earth? (2) Where is water readily available, both geographically and through time (e.g., seasonally)? (3) How fast will water taken from some source be replenished? These questions—linked to the endless cycling of water through air, land, and the oceans, referred to as the **hydrological cycle**—have fascinated people for a long time. In fact, much of the modern view of the functioning of the **hydrological cycle** originated several hundred years ago (Pfister et al., 2009).

1.2 The Sources and Locations of Fresh Water

How much water is on Earth? We typically think of the total water on Earth as existing in a variety of compartments or reservoirs: (1) the oceans, (2) icecaps and glaciers, (3) subsurface water (soil moisture and **groundwater**), (4) surface water (lakes, rivers, and wetlands), (5) the atmosphere, and (6) biological water (water in plants and animals). The oceans and the saline water stored in lakes and the ground contain about 1,350 million cubic kilometers, which is

Table 1.1 Sizes and residence times for major reservoirs in the hydrological cycle. Data from Maidment 1993.

Reservoir	Volume (km^3)	Percentage of total	Percentage of fresh water	Average residence time (years)
Water in:				
Land areas	47,960,240	3.5		4
Wetlands	11,470	0.0008	0.03	
Lakes:				
Fresh	91,000	0.007	0.3	
Saline	85,400	0.006		
Rivers	2,120	0.0002	0.006	
Subsurface				20,000
Soil moisture	16,500	0.001	0.05	
Groundwater:				
Fresh	10,530,000	0.8	30.1	
Saline	12,870,000	0.9		
Biological water	1,120	0.0001	0.003	
Icecaps and glaciers	24,364,100	1.8	69.6	
Atmospheric water	12,900	0.001	0.04	0.02
Oceans	1,338,000,000	96.5		2,650
Total	1,385,973,140	100	100	

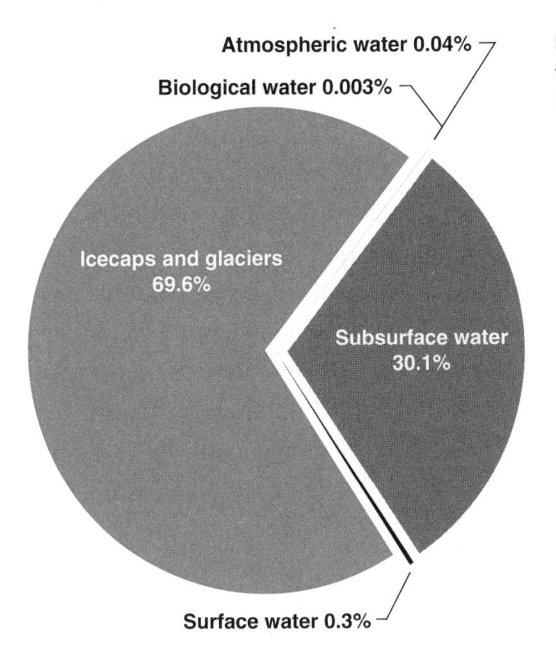

Figure 1.2 The distribution of Earth's fresh water. Data from Maidment 1993.

about 97.5% of the total amount of water on Earth. The approximate volumes of water in the other reservoirs listed make up the remaining 2.5% of the total water (Table 1.1). A large fraction of the fresh water is locked in snow and ice, and much of the liquid fresh water is below the surface of Earth, held in soils and rocks (Figure 1.2).

The absolute numbers associated with the volumes of water on Earth are huge and often hard to comprehend easily. One exercise to help envision the sizes of the different reservoirs is to think of a series of Olympic-sized swimming pools. Suppose that we shrink the reservoirs such that the world's saline water fills the first swimming pool fully to its 2-meter (~6.5-foot) depth. In the next pool, we place the equivalently shrunken total fresh water on Earth. This pool would be filled to a depth of 5 centimeters (~2 inches). If we eliminated all the frozen water and placed liquid fresh water in the next pool, the depth would be 1.5 cm (a little more than half an inch). For the last pool, if we restrict ourselves to the fresh water that we see at the surface in rivers and lakes, the depth would be about 1.5 millimeters (about half the thickness of a fingernail).

1.3 The Global Hydrological Cycle

To approach an answer to the question about where water is readily available, both geographically and through time, we must consider how water moves in the Earth system and what some of the controls are on the time and space variations of the processes. Water in the various compartments or reservoirs noted above can move within the reservoir itself—think of an ocean current such as the Gulf Stream carrying water from one place to another or a large river such as the Ganges moving water from the mountains to somewhere downstream. Water can also move from one reservoir to another. Water flows from high energy potential to low energy potential. For example, because elevation is one measure of potential energy, water may flow from high to low elevation. This endless movement of water around the globe is known as the **hydrological cycle**. Water exists in three states—solid, liquid, and gas—and can move within the global cycle in any of these states.

Because the focus of this text is on freshwater resources, let us first consider the process of precipitation that delivers water to the land surfaces on Earth (Figure 1.3). Frozen precipitation reaching the surface can be stored for relatively short time periods on the surface (think of snow in mid-latitudes, which begins to melt within days after an event), can be stored in seasonal snowpacks (think of ski areas), or can accumulate in long-lasting, "permanent" snowpacks (think of glaciers or of Antarctica). A portion

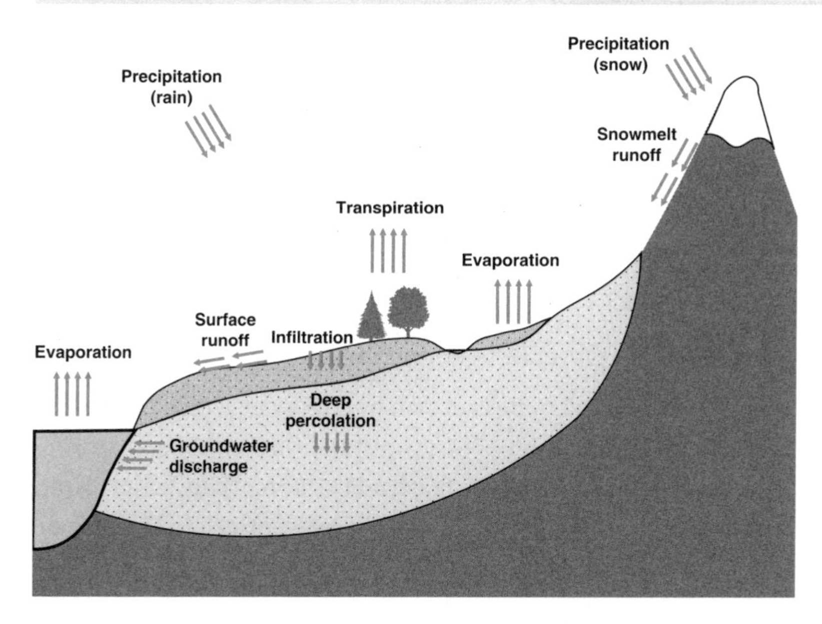

Figure 1.3 Mechanisms of water movement within the hydrological cycle. Water movement from the atmosphere to the oceans and continents occurs as precipitation, which includes rain, snow, sleet, and other forms. On the continents, water may be stored temporarily but eventually returns to the oceans through surface runoff and groundwater discharge or to the atmosphere through evaporation or transpiration. Figure redrawn from Hornberger et al. 2014.

of the unfrozen precipitation that reaches the surface as rain can be retained for short periods in depressions or on leaf surfaces in a process referred to as **interception**. Some precipitation infiltrates into soil (**infiltration**) or flows across the land surface (**surface runoff**) and collects into surface water bodies such as streams and rivers. Some of the water that stays on the surface evaporates back into the atmosphere (**evaporation**) in a "short circuit" of the global cycle.

Some of the surface water that flows in rivers—with possible multiple delays in passing through lakes and reservoirs—ultimately reaches the ocean to complete this arm of the global cycle. Some of the rain or snowmelt that enters the soil is retained near the surface and is returned to the atmosphere by plants. In a process referred to as **transpiration**, plants take up water through their roots, transport it upward through their stems, and release it to the atmosphere through openings in their leaves.

Part of the infiltrated water percolates vertically downward through the soil (**deep percolation**) and reaches a saturated part of the subsurface where it flows as **groundwater**. **Groundwater** flow also is driven by topography and pressure gradients, discharging to areas where it returns to the surface as the **baseflow** of streams and rivers, as freshwater springs, or as submarine discharge directly to the ocean. Water evaporates from the surface of the oceans and thereby replenishes the water in the atmosphere, which in turn replenishes water on the continents through precipitation.

1.4 Rainfall: Water Delivery by the Atmosphere

The **hydrological cycle** is driven by energy provided by the sun. The uneven distribution of heating at the surface of Earth creates differences in air temperature that lead to thermal convection, similar to the movement of heated water in a pan placed over a burner. Consider the effect of a very warm tropical sun at the equator. The moist air near the surface is heated strongly, causing it to expand and become less dense so it rises vertically upward (Figure 1.4); this is often referred to as a low-pressure zone. As the air rises, it is cooled. Moisture is more likely to condense, forming clouds, when the air is cool than when the air is warm. As the air rises, it cools, and this results in significant rainfall in the tropics.

The rising air ultimately reaches a "lid" on the lower atmosphere (the tropopause), and this now-drier air is diverted in flows to the north and the south. The air flowing away from the equator continues to cool, becoming more dense. At about 30 degrees N and 30 degrees S, the cool, dry air sinks to the surface and flows back toward the equator at the surface; this is often referred to as a

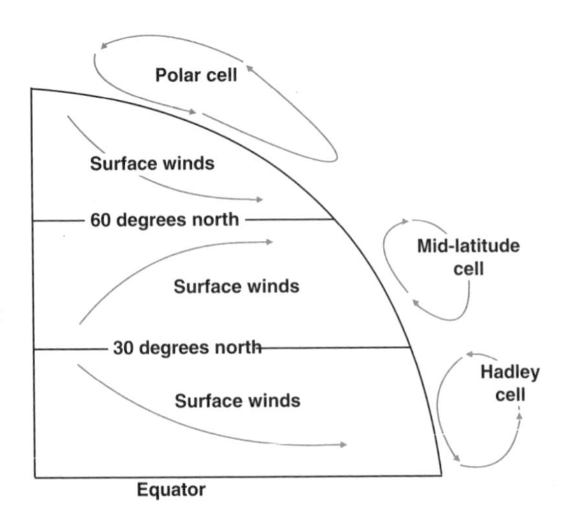

Figure 1.4 Hadley circulation. Warm, moist air at the equator rises, cools, and produces rainfall. The air then flows away from the equator, cooling as it moves. At about 30 degrees N and S latitude, the cool, dry air sinks to the surface, and the return circulation at the surface is toward the equator. The dynamics of the mid-latitude and polar cells are similar.

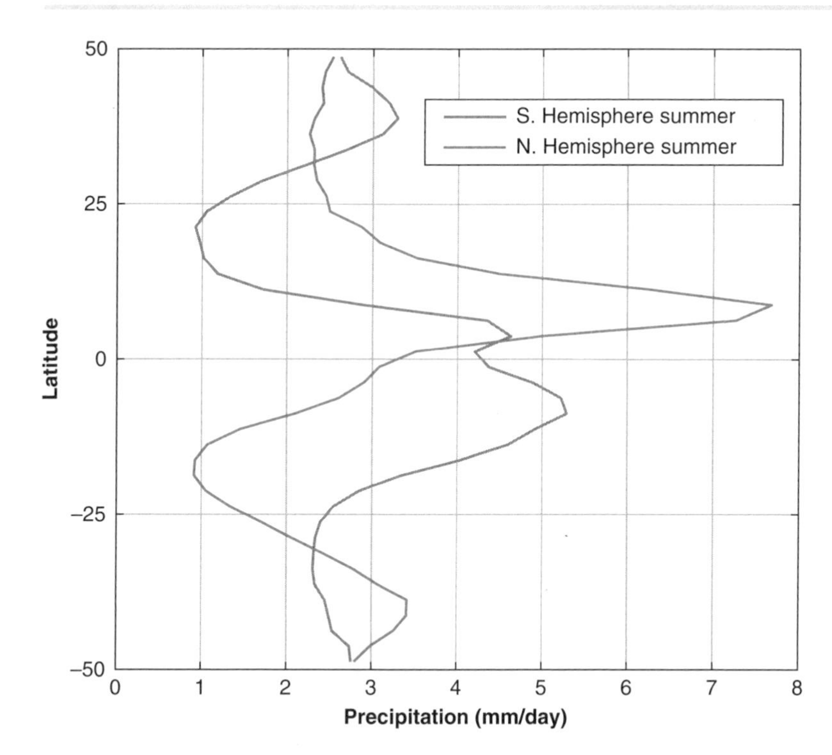

Figure 1.5 Average (1979–2008) seasonal precipitation as a function of latitude. Precipitation is greatest in the summer near the equator in the Northern and Southern Hemispheres.

high-pressure zone. This cyclic motion is known as **Hadley circulation**. Two other such cells, the mid-latitude cell that has rising air at about 60 degrees N and 60 degrees S and sinking air at about 30 degrees N and 30 degrees S and a polar cell with rising air at about 60 degrees N and 60 degrees S and sinking air at the poles, complete the picture of large-scale atmospheric motions.

The large-scale circulation has profound influences on available freshwater resources. The surface flows returning toward the equator from 30 degrees N and 30 degrees S pick up moisture by **evaporation** and converge on a zone near the equator known as the **intertropical convergence zone (ITCZ)**. The regions of Earth affected directly by the **ITCZ** generally receive copious amounts of precipitation whereas the zones where the cool, dry air subsides are generally deficient in precipitation. Consider that the major deserts of the world tend to be located near 30 degrees N and 30 degrees S. In short, the tropics receive significantly more precipitation than the subtropics or the poles (Figure 1.5).

The large-scale circulation of the Hadley and other cells describes some major features that determine relatively dry and relatively wet regions of the globe. Within the broad regions, however, there are processes that occur on smaller scales that also determine precipitation occurrence. Clouds form when moisture in relatively warm air at the surface rises, cools, and condenses. When conditions are right, the small water droplets in clouds coalesce into larger drops, and these can then fall to the ground as rain or snow. Generally, precipitation requires uplift and moist air. Convergence, as for the **ITCZ**, is one type of uplift.

Several other mechanisms can force air near the surface to rise. The air near the ground can be warmed by the sun, for example. On a sunny summer day, the warmer, lighter air can rise due to convection; this convective precipitation is the hallmark of summer thunderstorms. Moving air encountering a mountain chain also will be forced to rise; if there is moisture in the air, clouds will form, and rain (**orographic precipitation**) will occur on the windward side of mountains. On the leeward side of mountains, the air descends and is warmed, so clouds tend to evaporate. This is why areas downwind of high mountains often are arid or semiarid. A third major mechanism for forcing air at the surface to rise is the progression of weather fronts. A cold front, for example, is the boundary between a mass of relatively cold air that is pushing into a mass of relatively warm air. When this occurs, the warm air is forced to ride up over the wedge of cold air, again resulting in condensation and often precipitation.

The passage of seasons can have strong effects on precipitation regionally. The differential heating of the hemispheres, for example, causes the **ITCZ** to move north during the Northern Hemisphere summer and south during the winter. That is, the **ITCZ** and the Hadley cells shift toward the hemisphere that receives the most solar radiation, with a corresponding shift in precipitation patterns (see Figure 1.5). Correspondingly, there are large, semipermanent high-pressure systems, such as the "Bermuda high" in the Atlantic Ocean, that move north and south with the seasons and have a strong influence on the movement of humid air from the oceans to the continents.

1.5 The Land Surface: Dividing Water among Soils, Streams, and Aquifers

When rainfall reaches the surface of Earth, initially it may be held on the surfaces of vegetation or in depressions (e.g., puddles), but if enough water is delivered to fill these storages, ultimately the water either must infiltrate into the soil or must move across the surface, collecting in streams, rivers, or oceans. Water that is held at the surface—on leaf surfaces, in depressions, or in rivers and lakes—is subject to **evaporation**. How much of the precipitation goes to **infiltration** depends on the characteristics of the surface. If the surface is bare rock or covered by impervious areas, little or none of the water will infiltrate; instead, it will become **surface runoff**. If the surface is the soil in a temperate forest, essentially all the rainfall or melted snow will infiltrate. A fraction of the water that infiltrates is held in the pores of the soil and becomes available to plant roots and **transpiration**. Surface soils can hold only a certain amount of water, termed the **available water capacity** (AWC), and once that capacity is reached, water will percolate downward in the subsurface to an **aquifer**, becoming **groundwater**.

1.6 The Thornthwaite Water Budget

A widely used method for representing the nature of available water resources locally in an area is to look at the various processes that deliver water to and remove water from a landscape. Thornthwaite (1948) developed a method to calculate the water budget by accounting for rainfall, **evapotranspiration** (combined **evaporation** and **transpiration**, abbreviated as ET), and soil storage. The method is conceptually simple. The soil is treated as a bucket with a unit cross-sectional area, so "volume" is expressed as the depth of water. This idea is perhaps more intuitive if one thinks about how we describe rainfall—not as a volume but as a volume divided by area. So the total rainfall for a storm is expressed as a depth—for example, in millimeters. The size of the bucket is the **available water capacity** (AWC) of a soil. Every month, rain or snowmelt is added to the soil bucket, and **evapotranspiration** is removed. When the bucket is full (at the AWC), any additional rain for that month either percolates out of the bucket to **groundwater** or becomes **surface runoff**. The Thornthwaite budget does not distinguish among losses out of the bucket by percolation, **groundwater** runoff, or **surface runoff**.

To compute a water budget for a given location, the required data and information include rainfall, soil AWC, and **evapotranspiration**. Rainfall data are readily available from meteorological records for many areas, and global databases have been assembled (Willmott & Matsuura, 2001). Soil **AWC** values have been developed by many soil survey agencies, and global data sets are available (Global Soil Data Task Group, 2000). **Evapotranspiration** is not routinely measured, however, so it must be calculated by using other information.

The rate of **evapotranspiration** depends on atmospheric conditions and on whether water is available at the land surface. The Thornthwaite water budget approach uses the concept of **potential evapotranspiration (PET)**. **PET** is defined as the rate at which water would be transferred from a wet surface to the atmosphere under prevailing meteorological conditions. Loosely speaking, **PET** is the rate at which the atmosphere can remove water from Earth's surface, provided there is adequate water available. Actual **evapotranspiration** (AET) may equal **PET** over a well-watered lawn, but it will be much less than **PET** in the middle of a desert where the atmospheric conditions are excellent for removing water from the surface but there simply is no water to remove.

There are a variety of methods for calculating **PET** from meteorological data. Among the easiest to use are temperature index

Box 1.1 The Hamon Method

Compute, e_s, the saturation vapor pressure for each day:

$e_s = 0.6108\exp[(17.27T)/(237.3 + T)]$ where T is the temperature in Celsius.

Compute H_t, the number of daylight hours on day t:

$H_t = (24\omega_s)/\pi$
where ω_s is the sunset hour angle of day t:
$\omega_s = \arccos(-\tan\Phi \tan\delta)$,

where Φ is latitude and δ is the solar declination on day J (Julian day) of the year

$= 0.4093\sin[(2\pi/365)J - 1.405]$

Calculate PET using:

$$\text{PET} = \frac{(2.1 \times H_t^2 \times e_s)}{(T + 273.2)}$$

All angles are expressed in radians. Haith and Shoemaker (1987) set PET = 0 on days for which $T \leq 0$.

methods. These methods use temperature data and geographic location to compute **PET**. One temperature index model is the Hamon (1961) method (Box 1.1). Temperature methods are convenient because global temperature data are available at the same spatial resolution as precipitation (as mentioned earlier for precipitation data).

A water budget can be calculated using the following set of variables. Note that consistent units must be used.

- Monthly precipitation (P).
- **Potential evapotranspiration** (PET).
- Storage (ST): The amount of soil moisture (water) held in the soil at any particular time.
- Change (or "delta") in storage from 1 month to the next (DST).
- Actual evapotranspiration (AET): The amount of water delivered to the air from the processes of **evaporation** and **transpiration**, which depends on the moisture available, temperature, and humidity.
- Deficit (D): Soil moisture deficit that occurs when AET exceeds the PET, when the demand for water surpasses that which is actually available.
- Surplus (S): Soil moisture surplus that occurs when the PET exceeds the AET and the soil is at its **AWC**. Surplus water cannot be added to the soil when it is at **AWC**, so it runs off at the surface or percolates as recharge.

The Thornthwaite water budget is readily calculated using a spreadsheet. Precipitation data are entered on the first line under each month, and calculated values for **PET** are entered on the second line using consistent depth units (Table 1.2). The precipitation minus the potential evapotranspiration (P-PET) is calculated on the third row. A value for the **AWC** is chosen based on the location; for instance, for a location in the humid temperate zone of the Northern Hemisphere—such as Tennessee in the United States—an assumption that the soil storage (ST) is at the **AWC** for the soil in January is reasonable, so that value is placed in the January column for storage. Because the P-PET for January is positive, the DST must be zero. The actual ET for January equals the **PET**, and the soil deficit is zero. The surplus, either percolation to **groundwater** or **surface runoff**, is equal to the P-PET. The calculation proceeds from month to month. Note that in June for the Crossville location, the P is less than the **PET**, so water is taken from

Table 1.2 Thornthwaite water balance for Crossville, Tennessee, 2007. Precipitation and temperature (used in calculation of potential evapotranspiration) data from US National Oceanic and Atmospheric Administration (NOAA).

	JAN	FEB	MAR	APR	MAY	JUN	JUL	AUG	SEP	OCT	NOV	DEC
P	117.9	52.8	53.3	115.6	107.2	59.7	69.1	34.5	69.1	94.0	126.5	89.4
PET	4.8	0.0	39.6	37.1	90.8	121.8	127.0	150.0	97.5	57.4	17.5	12.7
P-PET	113.1	52.8	13.7	78.5	16.4	−62.1	−57.9	−115.5	−28.4	36.5	109.0	76.8
DST	0.0	0.0	0.0	0.0	0.0	−62.1	−12.9	0.0	0.0	36.5	38.5	0.0
ST	75.0	75.0	75.0	75.0	75.0	12.9	0.0	0.0	0.0	36.5	75.0	75.0
AE	4.8	0.0	39.6	37.1	90.8	121.8	81.9	34.5	69.1	57.4	17.5	12.7
D	0.0	0.0	0.0	0.0	0.0	0.0	45.1	115.5	28.4	0.0	0.0	0.0
S	113.1	52.8	13.7	78.5	16.4	0.0	0.0	0.0	0.0	0.0	70.5	76.8

Note: All values are in millimeters of water. AE, actual evapotranspiration; D, deficit; DST, change in storage for the month; P, monthly precipitation; PET, potential evapotranspiration; P-PET, precipitation minus potential evapotranspiration; S, surplus; ST, storage.

soil storage. In July, the soil storage is depleted, so the AET falls below the **PET**.

1.7 Average Residence Time of Water in Various Reservoirs

In planning for the use of water resources it is important to understand how much water is available and where and when it is available. It is clear from looking at water budgets for many places that there may be a need to find a way to use a surplus in, say, March, during a period of deficit in, say, August, for many locations in the Northern Hemisphere. For example, deficits in the water budget tend to occur in summer, exactly the period that is critical for food production. Humans withdraw water from surface water and **groundwater** reservoirs for various uses, including during times when P-PET is less than zero. These withdrawals ultimately must be replenished by precipitation or discharge if the water use is to be sustainable. The third question posed at the beginning of this chapter (How fast will water taken from some source be replenished?) is of importance to water resources managers.

One simple measure to address the replenishment time question is the **average residence time** (see Table 1.1). **Average residence time** is calculated by dividing the volume of a reservoir (see Table 1.1) by the average flow rate (Figure 1.6) through it. For example, if a pond in a river has a volume of 300 cubic meters (m^3) and the inflow rate to (and outflow rate from) the pond is 3 m^3/minute, on average water will reside in the pond for $300/3 = 100$ minutes. Of course,

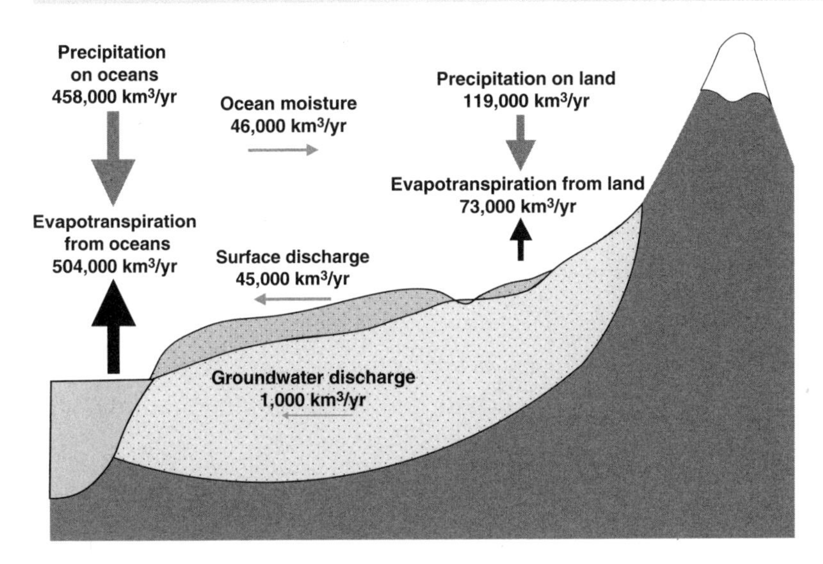

Figure 1.6 Flows within the hydrological cycle. Black arrows depict flows *to* the atmosphere, gray arrows depict flows *to* land or oceans, and blue arrows indicate lateral flows. Data from Maidment 1993.

this is an **average residence time**—some water flowing near the center of the river may move much more quickly through the pond than slower moving water near the banks. But the **average residence time** does give a reasonable index to answer the replenishment question. If the pond were emptied at some time, it would take 100 minutes for the inflow to fill it, assuming no outflow during that time.

Average residence times for the major Earth reservoirs vary widely. Water molecules spend on average only a little over a week in the atmosphere but more than 2,500 years in the ocean. For freshwater reservoirs, which are of greatest importance for water resources management, the residence times are on the order of a few months for seasonal snow and for rivers, about a month or two for soil moisture, several decades for large lakes, decades to centuries for shallow **groundwater**, and millennia for deep **groundwater**. Again, note that these are **average residence times**.

1.8 Concluding Remarks

The general circulation of water through the atmosphere, oceans, rivers, lakes, and beneath the ground determines the availability of freshwater resources at any location on Earth. As such, a basic understanding of the **hydrological cycle** is a prerequisite for any assessment of water resources. The calculation of water budgets at various time and spatial scales is fundamental to planning the use of water resources. These budgets ultimately must include the withdrawal and consumption of surface and **groundwater** by humans so that an understanding of the coupled natural–human system can be gained. Subsequent chapters of this book will explore these points in more detail.

1.9 Key Points

- Water is essential for life—Earth is a water planet. Humans use water for many purposes, including for agriculture, energy production, and basic needs such as drinking, cooking, bathing, and sanitation. (Section 1.1)
- Water can be considered to be held in a set of "reservoirs." The primary reservoirs are the atmosphere, the oceans, snow and ice, **groundwater**, and lakes and rivers. (Section 1.2)
- The global-scale, endless recirculatory process that links water in the atmosphere, on the continents, and in the oceans is known as the **hydrological cycle**. This cyclical process relates to the flows of water within and between various conceptual reservoirs—the atmosphere, oceans, and land. (Section 1.3)
- Precipitation occurs when moist air is forced to rise and cool, thereby causing water vapor to condense. Under favorable conditions, droplets coalesce and can fall to the ground as rain or snow. Rainfall patterns reflect both large-scale (global) and more regional atmospheric circulation. (Section 1.4)
- Water reaching the land surface in excess of small-scale surface storage is divided into **infiltration** and **surface runoff**. The infiltrated water is stored in the soil up to the **available water capacity (AWC)** and the excess percolates down to **groundwater**. (Section 1.5)
- Precipitation that reaches the land surface can be stored on vegetation or in depressions, can infiltrate into soils, or can flow over the surface as runoff. A monthly water budget is a useful way to represent how water is apportioned among the various paths. (Section 1.6)
- The **average residence time** of water in various reservoirs varies tremendously. Surface water and soil water residence times are typically expressed in months, but **groundwater** residence times are generally expressed as years or even millennia. (Section 1.7)

Table 1.3 Information for water balance calculation for Kuhlna, Bangladesh. Precipitation and temperature (used in calculation of potential evapotranspiration) data from Bangladesh Meteorology Department.

	JAN	FEB	MAR	APR	MAY	JUN	JUL	AUG	SEP	OCT	NOV	DEC
P	254.7	129.8	32.1	6.6	13.3	44.4	52.1	87.5	200.0	335.6	329.8	323.5
PET	186.5	157.7	93.6	42.9	34.6	61.7	136.6	199.9	222.6	220.6	202.7	197.8
P-PET												
DST												
ST	100.0											
AE												
D												
S												

Note: All values are in millimeters of water. AE, actual evapotranspiration; D, deficit; DST, change in storage for the month; P, monthly precipitation; PET, potential evapotranspiration; P-PET, precipitation minus potential evapotranspiration; S, surplus; ST, storage.

1.10 Example Problems

Problem 1.1. Consider the hypothetical example of an Olympic size swimming pool to help visualize the relative sizes of water reservoirs (see Table 1.1). Instead of taking the world ocean as filling the first pool, consider instead just the freshwater portion of total water. If the total fresh water fills the first pool (a depth of 2 m),

a. to what depth would a pool with just snow and ice be filled? (Hint: The depth will scale with the relative size of the reservoir because the lateral dimensions of the pool are fixed.)

b. to what depth would a pool with just subsurface water be filled?

c. to what depth would a pool with just lakes, wetlands, and rivers be filled?

Problem 1.2. For July at Big Meadows in Virginia, United States, the average number of daylight hours (H_t) is 14.4, the Julian day (J) at midmonth is 197, and the average temperature (T) is 18.9°C. Calculate the **PET** using the Hamon method. The precipitation for one particular July is 94.7 mm. Estimate how the amount of water stored in the soil would change during that month.

Problem 1.3. Given the information in Table 1.3 (all values in millimeters of water), complete the water budget calculation for Kuhlna, Bangladesh. Assume that the soil is at the **AWC** of 100 mm at the end of the rainy season in September.

Problem 1.4. The water budget for Big Meadows, Virginia, in the United States using data for the year 1964, which was a drought year, is shown in Table 1.4. For this calculation the **AWC** was taken as 60 mm.

Table 1.4 A Thornthwaite water balance for Big Meadows, Virginia, for 1964. Precipitation and temperature (used in calculation of potential evapotranspiration) data from US National Oceanic and Atmospheric Administration (NOAA).

	JAN	FEB	MAR	APR	MAY	JUN	JUL	AUG	SEP	OCT	NOV	DEC
P	126.0	121.2	65.6	157.3	47.9	36.1	83.9	63.8	99.8	96.0	114.0	85.7
PET	0.0	0.0	24.6	42.1	73.3	97.9	102.9	82.9	56.8	29.9	21.8	13.3
P-PET	126.0	121.2	41.0	115.2	−25.5	−61.8	−19.0	−19.1	43.0	66.1	92.2	72.4
DST	0.0	0.0	0.0	0.0	−25.5	−34.5	0.0	0.0	43.0	17.0	0.0	0.0
ST	60.0	60.0	60.0	60.0	34.5	0.0	0.0	0.0	43.0	60.0	60.0	60.0
AE	0.0	0.0	24.6	42.1	73.3	70.6	83.9	63.8	56.8	29.9	21.8	13.3
D	0.0	0.0	0.0	0.0	0.0	27.3	19.0	19.1	0.0	0.0	0.0	0.0
S	126.0	121.2	41.0	115.2	0.0	0.0	0.0	0.0	0.0	49.1	92.2	72.4

Note: All values are in millimeters of water. AE, actual evapotranspiration; D, deficit; DST, change in storage for the month; P, monthly precipitation; PET, potential evapotranspiration; P-PET, precipitation minus potential evapotranspiration; S, surplus; ST, storage.

a. The **AWC** always has to be estimated from some knowledge about soils, and there always is some uncertainty about what value of this parameter to use. Explore how changing the **AWC** from 60 mm to 90 mm changes the calculated water budget. Explain the results based on hydrological processes.

b. Climate change projections for Virginia indicate an expected warming of 2°C (3.6°F) and slightly increased precipitation, perhaps by about 5%. Given the parameters used in the Hamon **PET** calculation in Table 1.5, calculate how the water balance for a drought year like 1964 would be affected by these climate changes. Assume the **AWC** is 60 mm.

Table 1.5 Data for Hamon calculation of potential evapotranspiration. Temperature data from US National Oceanic and Atmospheric Administration (NOAA).

	JAN	FEB	MAR	APR	MAY	JUN	JUL	AUG	SEP	OCT	NOV	DEC
T (°C)	−1.4	−3.6	2.9	8.4	14.2	18.3	19.2	17.7	15.2	8.0	6.9	0.8
J	15.5	46.0	76.5	107.5	137.0	166.5	197.0	227.5	258.0	288.5	319.5	350.0
δ	−0.4	−0.2	0.0	0.2	0.3	0.4	0.4	0.2	0.0	−0.2	−0.3	−0.4
Ω_S	1.3	1.4	1.5	1.7	1.9	1.9	1.9	1.8	1.6	1.4	1.3	1.2
H_t	9.6	10.5	11.8	13.1	14.1	14.7	14.4	13.5	12.3	11.0	9.9	9.3

Note: Big Meadows is located at latitude 38.5 degrees (0.67 radians) north.

Problem 1.5. A lake, with straight vertical sides, is at steady state. The lake is 4 m deep and 0.1 km^2 in area. Precipitation and runoff to the lake are 45,000 m^3 per year. Calculate the residence time of the lake in days.

1.11 Suggested Readings

Oki, T., & Kanae, S. (2006). Global hydrological cycles and world water resources. *Science*, *313*, 1068–1072.

University of Oregon. (2016). Global climate animations: Global water balance. http://geography.uoregon.edu/envchange/clim_animations/

1.12 References

Global Soil Data Task Group. (2000). Global gridded surfaces of selected soil characteristics (IGBP-DIS). ORNL Distributed Active Archive Center. https://doi.org/10.3334/ornldaac/569

Haith, D. A., & Shoemaker, L. L. (1987). Generalized watershed loading functions for stream flow nutrients. *Water Resources Bulletin, 23,* 471–478.

Hamon, W. R. (1961). Estimating potential evapotranspiration. *Journal of the Hydraulic Division, 87,* 107–120.

Hornberger, G. M., Wiberg, P. L., Raffensperger, J. P., & D'Odorico, P. (2014). *Elements of physical hydrology* (2nd ed.). Baltimore: Johns Hopkins University Press.

Maidment, D. R. (Ed.). (1993). *Handbook of hydrology*. New York: McGraw-Hill.

NASA. (2011). The water planet. http://www.nasa.gov/multimedia/imagegallery/image_feature_1925.html

Pfister, L., Hubert, H., & Savenije, G. (2009). *Leonardo Da Vinci's water theory: On the origin and fate of water*. Wallingford, United Kingdom: IAHS Press.

Solomon, S. (2010). *Water: The epic struggle for wealth, power, and civilization*. New York: HarperCollins.

Thornthwaite, C. W. (1948). An approach toward a rational classification of climate. *Geographic Review, 38,* 55–94.

Willmott, C. J., & Matsuura, K. (2001). Terrestrial air temperature and precipitation: Monthly and annual time series (1950–1999). NOAA data version 4.01. http://www.esrl.noaa.gov/psd/data/gridded/data.UDel_AirT_Precip.html

CHAPTER 2

Surface Water Resources

2.1 Introduction

The history of civilization is inextricably linked with water at the surface of Earth. Highly successful early civilizations arose along rivers that provided fertile floodplains and water for irrigation, navigation, and personal use. In time, rivers were harnessed to provide power, and industries grew up along many major river corridors. Water is withdrawn from lakes, wetlands, rivers, and human-made lakes (reservoirs) created by dams on rivers for many purposes, including most prominently for public water supplies, irrigation, and industrial use, including for power production. According to the Food and Agriculture Organization (FAO) of the United Nations, in 2008 about 81% of all fresh water withdrawn for human use across the globe was from surface water sources. In the United States, the US Geological Survey reported that 75% of the 423 cubic kilometers (km^3) of fresh water withdrawn in 2010 for various uses was from surface water sources. Clearly, water resource planners, managers, scientists, and engineers must have a basic understanding of surface water volumes and flows to carry out their work effectively.

2.2 Lakes

Lakes and wetlands contain 0.3% of the global fresh water (see Table 1.1), of which about 90% is found in freshwater lakes. Despite what seems at first glance to be a small fraction relative to the total fresh water on Earth, the water in lakes represents the lion's share of the available, liquid fresh water on the Earth's surface. There are about 18,000 natural lakes globally, with surface areas of at least 10 kilometers squared (km^2) (Downing & Duarte, 2010); a large portion of the volume of water is stored in a relatively small number of large lakes (Figure 2.1).

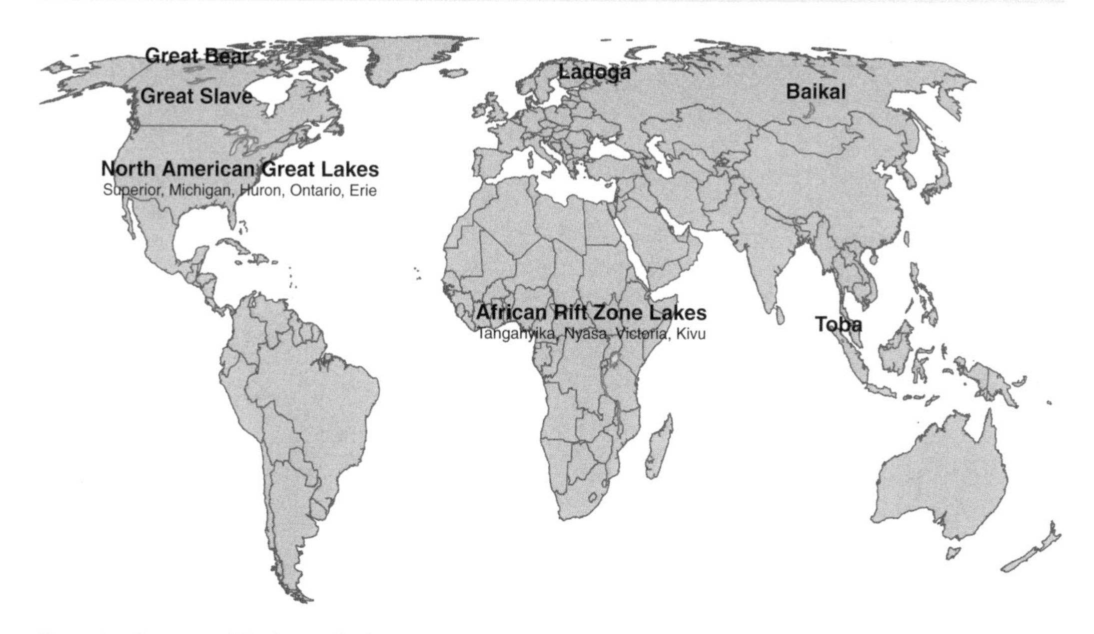

Figure 2.1 Fourteen of the largest freshwater lakes in the world by volume.

Lakes are formed by a variety of geological processes such as scouring by ice, water, or wind; formation of volcanic caldera; and tectonic shifts that create depressions along major fault lines. The volumes contained in lakes vary greatly (Figure 2.2). Because many natural lakes were formed by processes associated with glaciation, natural lakes are more prevalent at high latitudes than elsewhere. Lakes also support a diversity of ecosystem types and can be direct sources for water withdrawals for large populations (e.g., along the Great Lakes in the United States and Canada). Nevertheless, when compared with rivers, lakes do not tend to be as ubiquitous a source of fresh water for humans, because many lakes are found in areas that are not especially conducive to agriculture or urban development.

Wetlands are a fairly broad category of surface water. Wetlands can encompass large areas that are perennially flooded, but they also include land at the margins of surface water bodies that are periodically inundated. Thus, terms such as swamps, marshes, fens, and bogs are all used to refer to wetlands. Wetlands fill an important niche in providing what are called **ecosystem services**, which refers to the benefits that are provided to humans by ecosystems, either directly or indirectly (Chapter 8). For example,

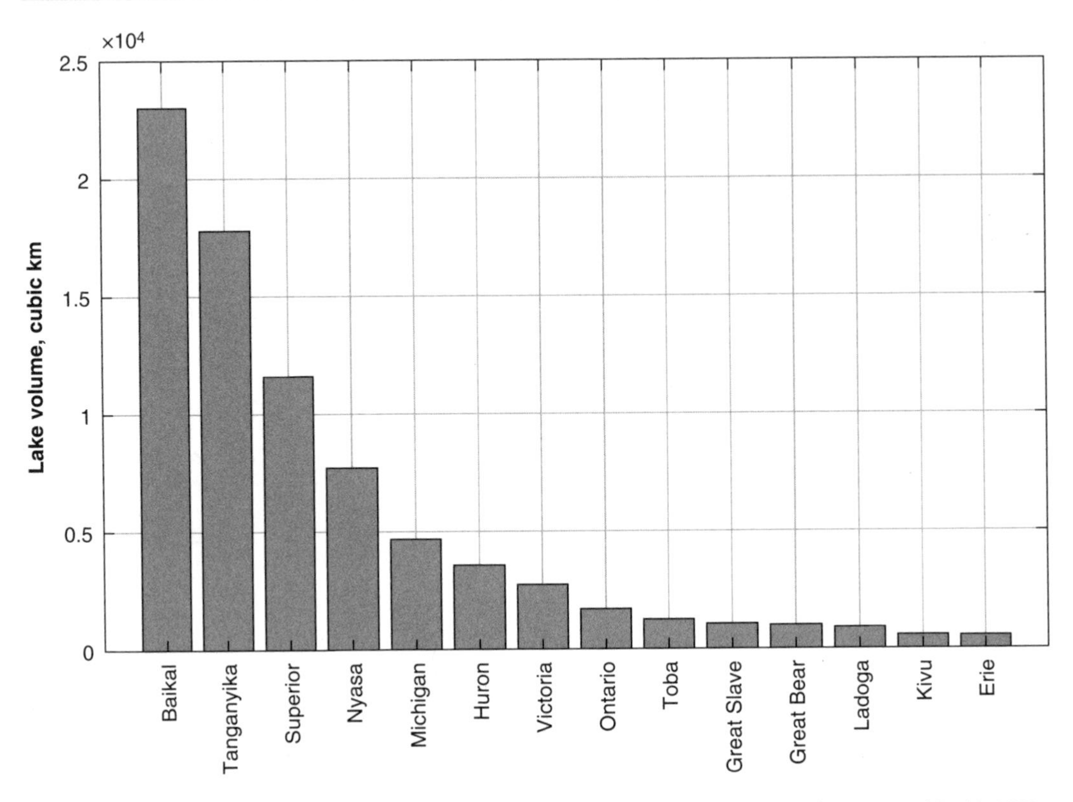

Figure 2.2 Freshwater lakes with volumes greater than 500 cubic kilometers. Data from Shiklomanov and Rodda 2003.

some wetlands act to mitigate river flooding by storing excess water and releasing it slowly. Wetlands have also been referred to as nature's kidneys, because they receive upstream water containing a variety of natural and **anthropogenic** constituents and remove them through biogeochemical processes. Wetlands also accumulate peat; the fossil fuels that are central to modern life arose through the burial of carbon in the geologically distant past, so peat is important with respect to our current attempts to limit global climate change. And, of course, wetlands are incredibly rich ecosystems in terms of biodiversity.

Humans have had an enormous impact on wetlands. Historically, wetlands were seen as unproductive or unusable, and they have been drained and filled for centuries. Estimates have placed the loss of inland wetlands at above 50% over the past two centuries (Davidson, 2014). Freshwater wetlands currently occupy about 5 million km^2, suggesting that at least 5 million km^2 of wetlands have been lost. Given the known importance of wetlands as habitat for many species of plants, birds, and other animals and their vital role in the provision of **ecosystem services**, the success of the ongoing efforts to stem and reverse the decline of wetlands is imperative.

2.3 Rivers

If the 0.3% of fresh water on Earth contained in lakes and wetlands seems small, the portion in rivers, 0.05%, is truly tiny. Yet the continuously resupplied flow of water in river systems is an incredibly important resource. Small streams drain small headwater catchments, and these small streams coalesce to become larger streams, which in turn coalesce to form large rivers that **discharge** to the world oceans at a variety of places (Figure 2.3). But not all rivers are equal. Even among the largest rivers, there is a wide variation in average annual volumes discharged over a given time period (Figure 2.4). Despite the small amount of water in rivers relative to the total available fresh water on Earth, these waters are overwhelmingly important.

To analyze the use of river water in terms of resource use, the basic quantity that needs to be determined is **discharge**, given as the volume flowing past a point on a river per time interval. For

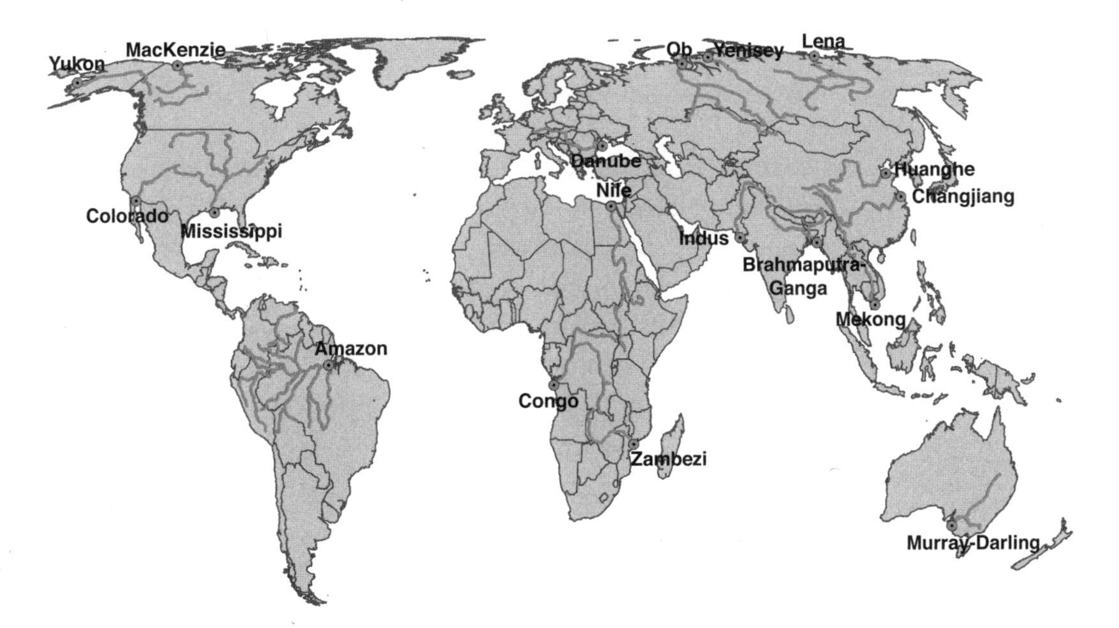

Figure 2.3 Points of discharge (circles) of 19 of the largest rivers (lines) in the world.

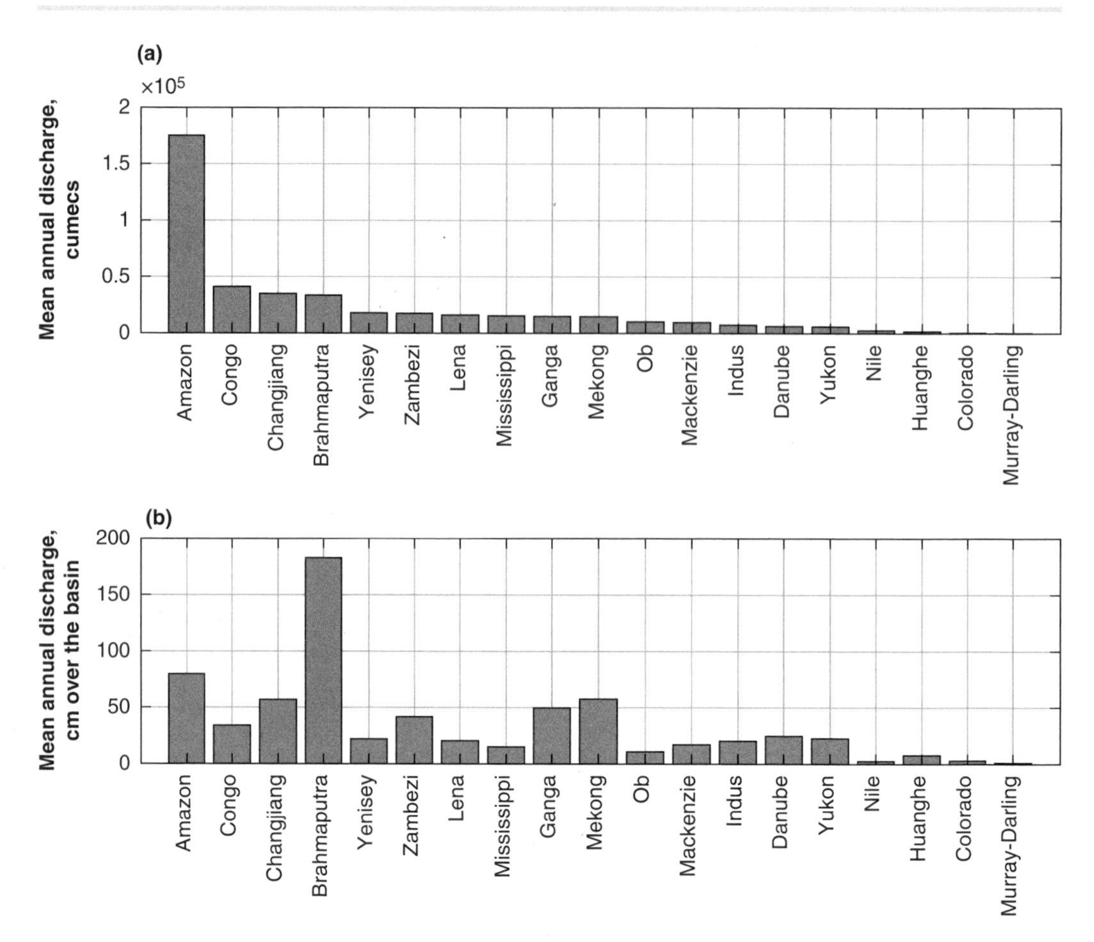

Figure 2.4 Average annual water discharge from the world's largest rivers expressed as cumecs (a) and depth of water over the basin area (b). Data from Shiklomanov and Rodda 2003.

example, **discharge** can be expressed in units of cubic meters per second (cumecs) or cubic kilometers per year (km^3/year) (see Figure 2.4a). By this measure, the large rivers of the world range from an annual **discharge** of 175,000 cumecs from the Amazon River in South America to fewer than 400 cumecs downstream of the confluence of the Murray and Darling rivers in Australia. The amount of water flowing in a river at a given point is conditioned by the area upstream that feeds water to it as well as by climate variables. This **catchment area** or basin area is defined by topographic divides that cause water to drain into different streams. The drainage basin area for an upland headwater stream may be only a few square kilometers whereas the drainage basin of a large river may be thousands of square kilometers.

To compare different rivers, it is convenient to express the flow rate normalized by area. For example, km^3/year divided by the basin area in km^2 gives km/year, the **discharge** expressed as an equivalent depth of water over the entire basin area over a selected time interval. Depth in kilometers is not very convenient itself, so the result could be expressed in centimeters or millimeters, the same units as used to express precipitation on the land surface. In this view of **discharge** normalized by basin area, the world's largest rivers range from about 180 cm/year for the Brahmaputra River, which drains to the Indian Ocean in Bangladesh, to about 1 cm/year for discharge from the Murray-Darling Basin, which drains to the Southern Ocean in Australia (see Figure 2.4b).

Most often, **discharge** is not measured directly; instead, the water level in a river, referred to as the **river stage**, is measured, and **discharge** is inferred from the stage measurement. The stage is recorded continuously using a pressure sensor, and **discharge** is measured occasionally using a current meter. The relatively infrequent concurrent measurements of both stage and **discharge** are used to form a **rating curve**, which is then used to calculate continuous **discharge** from the continuous stage measurements. A graph of **river stage** or **discharge** versus time at a point is called a **hydrograph**. The former is referred to as a **stage hydrograph** and the latter as a **discharge hydrograph** (Figure 2.5). By recording the stage at a given point in a river continuously, a continuous **discharge hydrograph** (Figure 2.6) can be established, enabling water resource planners, managers, and engineers to assess water availability at that location.

Much of the water in rivers and streams is used many times over in its transit from headwaters to **discharge** into an ocean or inland sea. That is, water withdrawals are used for many different purposes, and some water is used and returned to the river—typically somewhat the worse in terms of water quality. A portion of the water withdrawn becomes **consumptive use**, however, and is taken up by the atmosphere (through **evaporation** and **transpiration**) and not returned to the river.

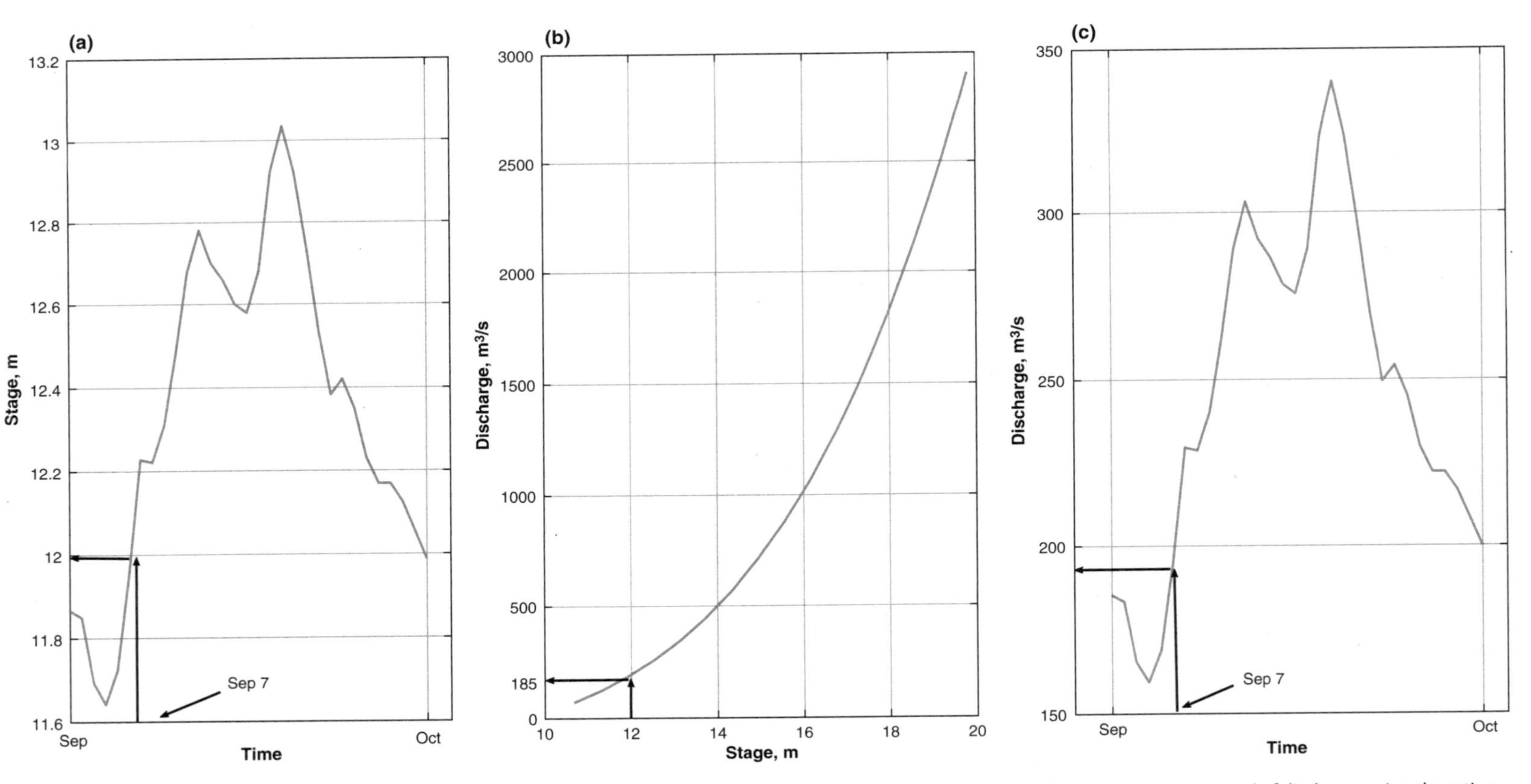

Figure 2.5 A stage hydrograph (a), a rating curve (b), and a discharge hydrograph (c). A continuous record of stage is converted into a continuous record of discharge using the rating curve. For example, on September 7, the stage is recorded as 12 m above mean sea level (arrows in panel a). The stage of 12 m corresponds to a discharge of 185 m^3/second (arrows in panel b). The value of discharge corresponding to the 12-m stage is assigned to September 7 for the discharge hydrograph (arrows in panel c). Data from the US Geological Survey.

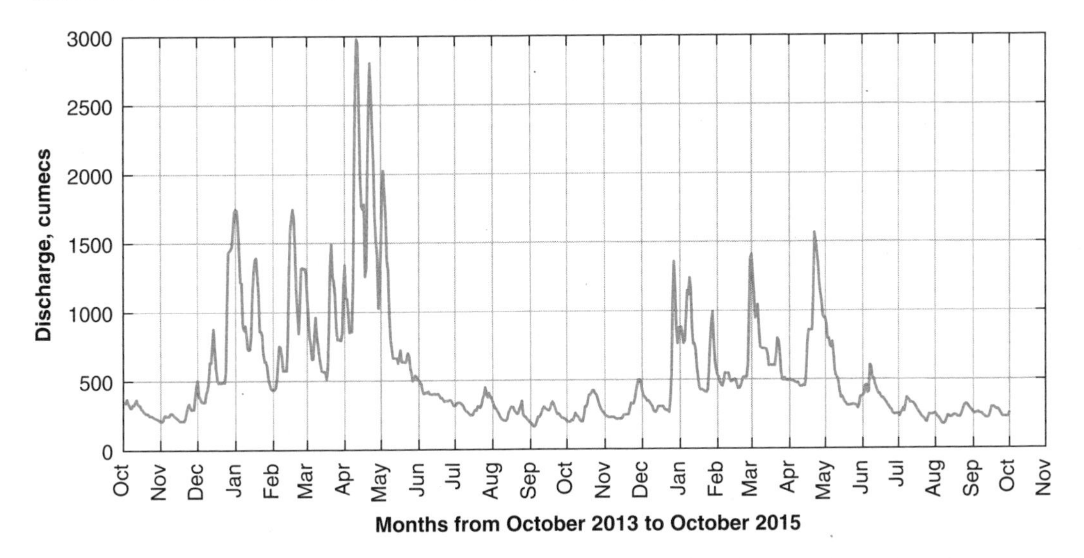

Figure 2.6 A discharge hydrograph for the Apalachicola River at Chattahoochee, Florida. During periods when evapotranspiration is low (from late autumn through early spring) rainstorms result in large peaks in the discharge hydrograph. During periods when evapotranspiration is high (late spring through early autumn) rainstorms result in relatively smaller peaks in river discharge. Data from the US Geological Survey.

2.4 Spatial and Temporal Distribution of Surface-Water Fluxes

The quintessential problem for those dealing with water resources is that water is not available uniformly, either by place or through time. That is, we often must deal with having either too much water (e.g., **waterlogging** or flooding) or too little water (e.g., drought). The flux of water—that is, the **discharge** per area per time—from one storage compartment (e.g., the land surface) to another storage compartment (e.g., the ocean) gives an indication of how water moves through regions of Earth.

The **discharge** or runoff of water from the continents to the oceans varies greatly from place to place. At the continental scale, average annual runoff ranges from 12.3 cm per year for Africa to 69.3 cm for South America (Table 2.1). **Discharge** also varies with latitude (Figure 2.7), as would be expected from our consideration of the atmospheric **Hadley circulation** in Chapter 1. And even within latitude bands, we know that there are relatively wetter and drier areas of Earth and that river flows reflect these differences. For example, the Gila River near Gila, New Mexico, in the United States drains 4,828 km^2 with an average annual flow of 4.85 cumecs (3.2 cm/year) and the Chattahoochee River at Atlanta, Georgia,

Table 2.1 Runoff from continents.
Data from Clark et al. 2015.

Continent	Total average annual discharge	
	cumecs	cm/year
Africa	4,000	13
Eurasia	16,300	29
South America	13,300	74
North America	7,100	29
Australia–Oceania	3,500	44

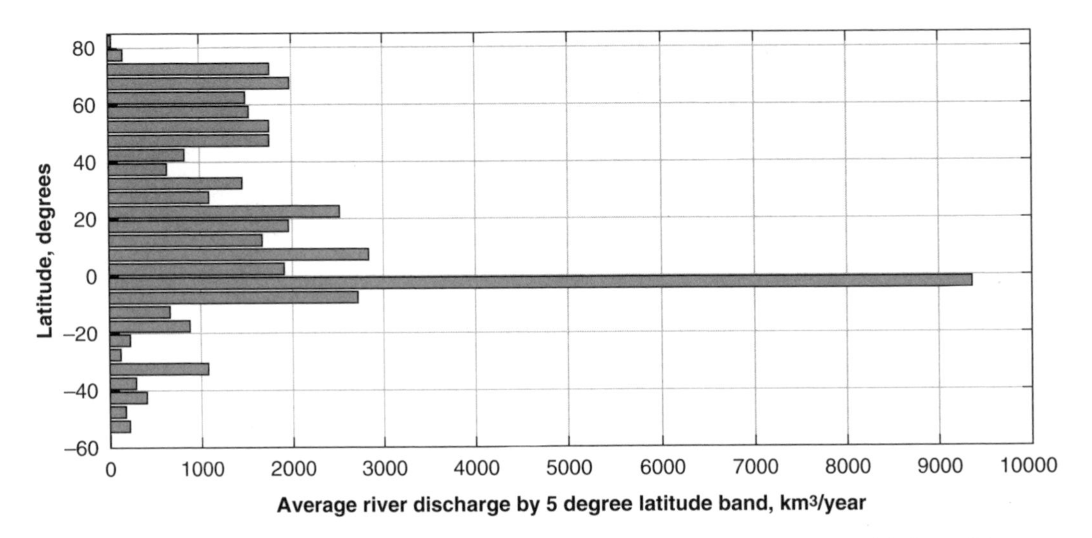

Figure 2.7 Average annual river discharge in km³/year by 5 degree latitude band. Data from the Global Runoff Data Centre 2014.

drains 3,756 km^2 with an average annual flow of 69.1 cumecs (22.8 cm/year). Both basins are between 33° and 34° N latitude, yet there is an order of magnitude difference in the average annual **discharge**.

River flows also vary tremendously with time. Rivers flood during times of extremely wet weather, and their flows diminish to very low levels during drought conditions. Some variations in time recur regularly because of seasonal effects. The impact of increases in **evapotranspiration** in summer months relative to winter months can lead to patterns of flow featuring relatively high river **discharge** every winter and relatively low flows every summer, even when rainfall may occur relatively evenly over the entire year as it does in many places in the southeastern United States (see Figure 2.6). In some areas, rainfall occurs seasonally and drives patterns of river **discharge**. For example, in Bangladesh the flow in the Ganges and

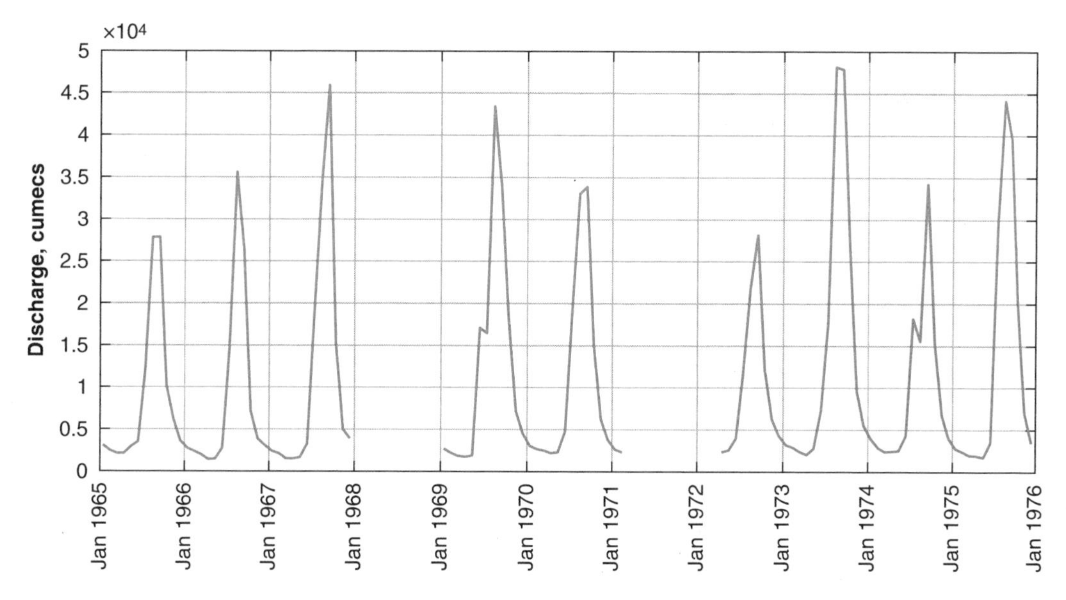

Figure 2.8 A discharge hydrograph for the Ganges River at Hardinge Bridge. The peaks in the discharge hydrograph reflect the seasonal nature of rainfall from the monsoon system, with the large bulk of rainfall occurring from July to October with dry conditions in other months. Data from Vorosmarty et al. 1998.

Brahmaputra rivers reflects the strong monsoonal system: about a third of the country is flooded each rainy season, with flows dropping drastically during the dry season (Figure 2.8).

Some variability in river flows is unanticipated in the sense that it occurs more or less randomly. We cannot predict either extreme floods or extreme droughts years in advance, nor do they predictably occur in any particular month or months. What can be done in terms of water resources planning is to use the statistics of recorded flows. For flood analysis, the statistic most often used is the **annual flood**, which is the highest value of **discharge** recorded in any year. In the United States, the **water year**, which begins on October 1 and ends on September 30 of the following year, is used. As an example, consider the Harpeth River at Franklin, Tennessee, in the United States. The data for water years 1974 through 2014 give us a record of 41 years. The 41 values of the highest daily **discharge** for each year comprise the **annual flood** series. The term "**annual flood**" is perhaps a misnomer because hydrologists refer to the values as annual floods regardless of whether the river actually overflowed its channel in a given year.

The annual floods can be ranked according to their size. For the Harpeth, the largest flood occurred in 2010, which is ranked 1, and the lowest flood flow occurred in 2012, which is ranked 41 (Table 2.2). Once we have the ranked annual floods, we can calculate the probability that a given **annual flood** will be equaled or exceeded in any given year (Equation 2.1). The probability, *P*, is the reciprocal of the recurrence interval *T*, of any given **annual flood**, *i*:

$$T_i = \frac{1+n}{R_i}$$
$$P_i = \frac{1}{T_i} \tag{2.1}$$

where n is the total number of years in the record and R_i is the magnitude ranking of the **annual flood** of interest.

Logarithms of flood flows and their probabilities are sometimes close to being normally distributed (i.e., approximating a bell-shaped curve). Because the logarithms are nearly normal, the flood flows can be plotted on a probability graph to estimate how likely it is for a flood of a given size to occur (Figure 2.9). The flood flows for the Harpeth River were ranked so that the scale on the vertical axis of Figure 2.9 is the probability that a flood associated with the value will be equaled or exceeded in any given year; it represents an **exceedance probability**. From the flood frequency diagram, a flood **discharge** of about 470 cumecs is associated with an **exceedance probability** of 0.05 per year. From this probability, we infer that there is a 5% chance in any year that a flood of that magnitude or larger will occur. The reciprocal of the **exceedance probability** gives the return period. The return period is the average interval between floods of the respective **discharge** or larger. For the Harpeth, 470 cumecs is the 20-year flood. Again, the 20-year flood does not mean that a flood of 470 cumecs will occur every 20 years for the Harpeth. The 20-year flood is a statistical

Table 2.2 Annual flood series, Harpeth River, Tennessee. Data from the US Geological Survey.

Rank	Maximum annual discharge (cumecs)	Year
1	507	2010
2	453	1975
3	343	1990
4	323	2004
...	...	...
41	61	2012

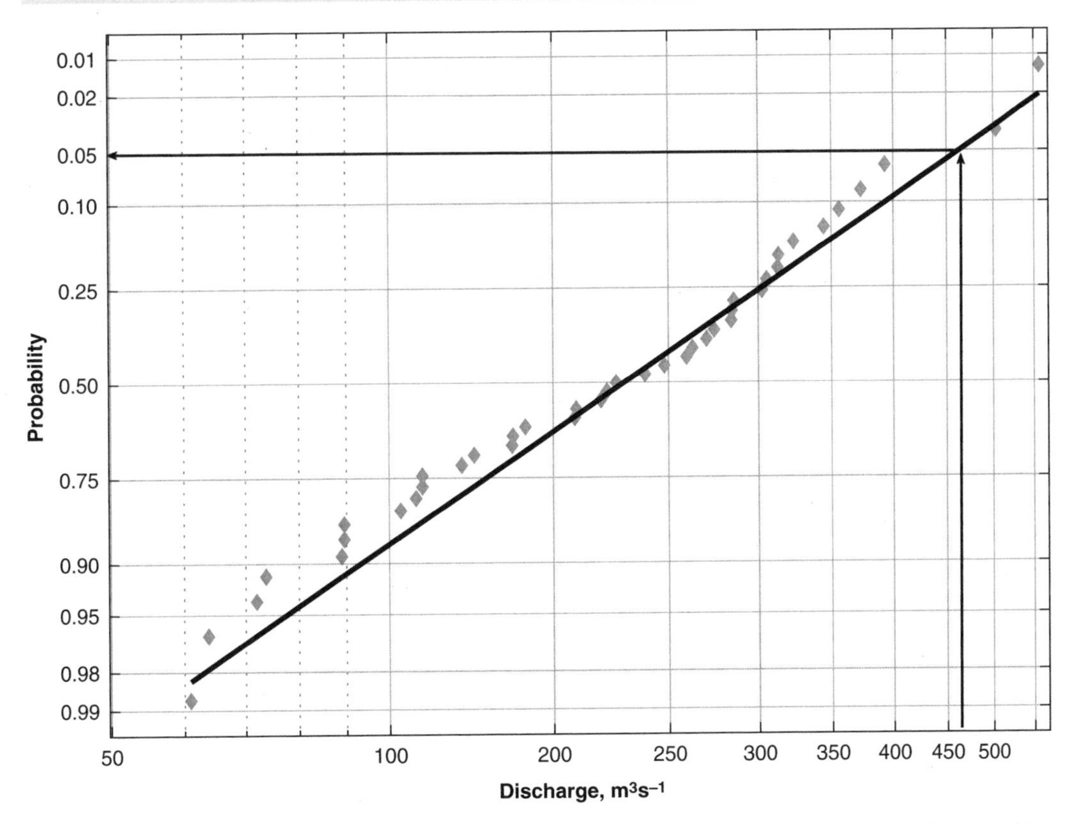

Figure 2.9 Flood frequency for the Harpeth River at Franklin, Tennessee. The arrows show an example, a discharge of 470 cumecs has an exceedance probability (P_i) of 0.05 so its recurrence interval (T_i) is 20 years. Data from the US Geological Survey.

estimate using historical **discharge** information; the 20-year flood suggests that there is a 1 in 20 or 5% chance each year of a **discharge** of 470 cumecs or greater to occur.

Development of land in flood-prone areas exposes buildings and people to a serious hazard. Because the damages that occur can be catastrophic, flood insurance is often recommended or even required for property owners. In the United States, the Federal Emergency Management Administration (FEMA) identifies the areas subject to a 100-year flood as special flood hazard areas. Recall that the 100-year flood is one that is expected to be equaled or exceeded *on average* once every 100 years—that is, the probability of such a flood occurring in any year is 0.01. The probability of a 100-year flood *not* being equaled or exceeded in a given year is 0.99.

Now suppose that you purchase an ice-cream shop that is in the 50-year floodplain and anticipate running it for 20 years. The

exceedance probability is 1/50 or 0.02. The probability that a flood will not occur in any year is 0.98, so this may seem relatively safe. But the probability that a 50-year flood will not occur in a 2-year period is 0.98 * 0.98, in a 3-year period is 0.98 * 0.98 * 0.98, and so forth. So the probability that your ice-cream shop will be flooded during the 20 years you plan to operate it is $1-(0.98)^{20}=0.33$. There is a one in three chance that you will sustain damage. Flood insurance may be a good option to consider.

Another way to visualize the temporal variability of flows in a river is the **flow duration curve**. All the daily flows are sorted by **discharge** and plotted against the fraction of days in the record when the flow was above a given value (Figure 2.10). The **flow duration curve** provides information for a first rapid assessment for water resources planning. If, for example, planners in the Harpeth River area were concerned about flood flows above 100 cumecs, the **flow duration curve** (see Figure 2.10) indicates that protection would be needed for about 1% of all days.

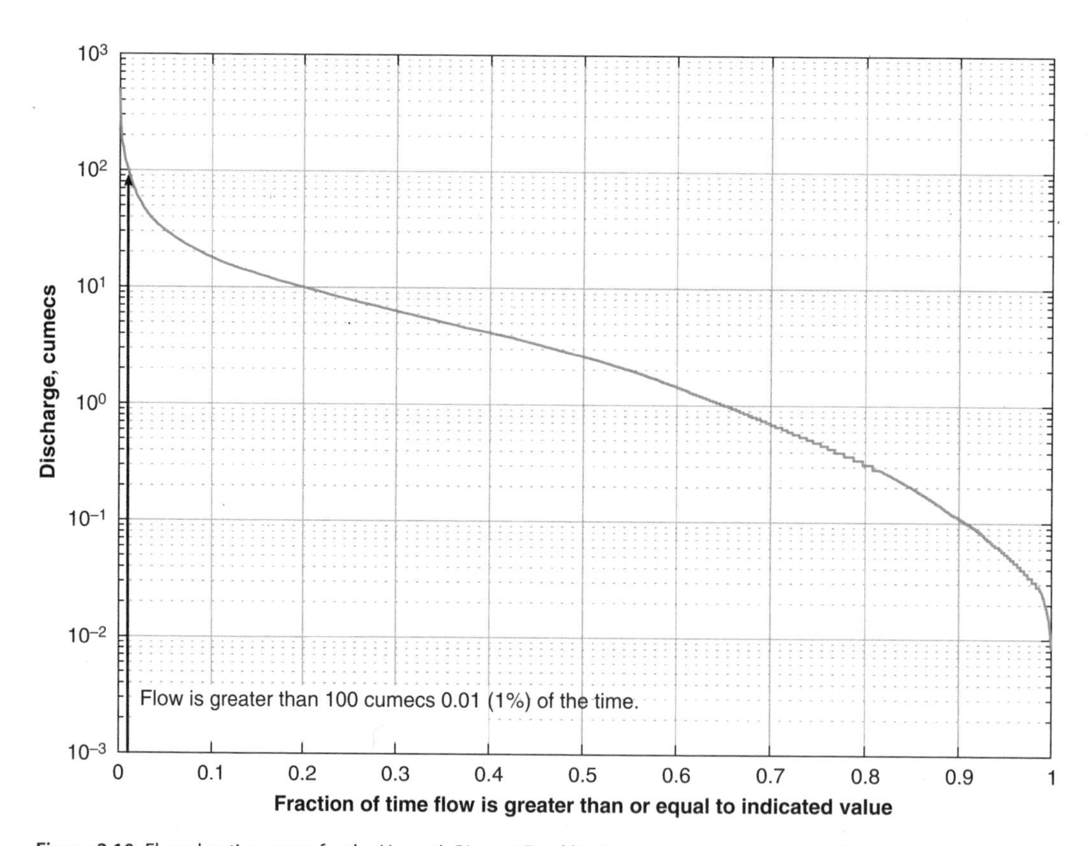

Figure 2.10 Flow duration curve for the Harpeth River at Franklin, Tennessee. Data from the US Geological Survey.

Conversely, the low-flow end of the **flow duration curve** gives an indication about the reliability of a river as a water supply. Franklin, Tennessee, is adjacent to the Harpeth River and draws water directly from the river for their municipal water supply. In 1950, the population of Franklin was about 5,500. In Tennessee, the average public supply per capita, which includes some industrial use as well as domestic use, is about 550 liters/day. For 5,500 people this indicates a daily requirement of 3 million liters, which converts to 0.035 cumecs. The **flow duration curve** indicates that this amount of water could be supplied reliably for about 97.5% of the time. By 2010, the population of Franklin had grown to 62,000, so the demand would have grown to about 0.4 cumecs. The flow in the Harpeth is above this level only about 74% of the time, so for more than 1 day in 4 the water supply would be inadequate to meet the demand. Clearly the Harpeth River itself is not adequate to supply the needs of Franklin today. Cities have a range of options to meet demands reliably, with one being the construction of dams to store water in reservoirs during high flow periods so that the water can be used when flows in the river drop.

2.5 Human-Made Lakes (Also Known as Reservoirs)

Reservoirs are human-made lakes. These lakes are developed for a variety of uses, such as storage for reliable water supplies for municipal and agricultural use, for hydroelectric power generation, for control of floods, and for recreation. The construction of dams worldwide accelerated greatly during the twentieth century (Figure 2.11). Although the pace of construction appears to have slowed a bit, there are many very large projects, especially for hydropower and irrigation, planned or underway that will continue impoundment of an ever-larger fraction of annual river flow. The Food and Agriculture Organization lists a significant number of dams with construction underway that have yet to be completed (FAO, 2015).

The benefits of reservoirs accrue from the services provided as planned in the construction—be they hydropower, irrigation supply, flood control, municipal water supply, or low-flow maintenance in a river during critical times of the year. The construction of dams has potentially negative impacts as well. The necessary inundation of large areas has a direct effect on people and

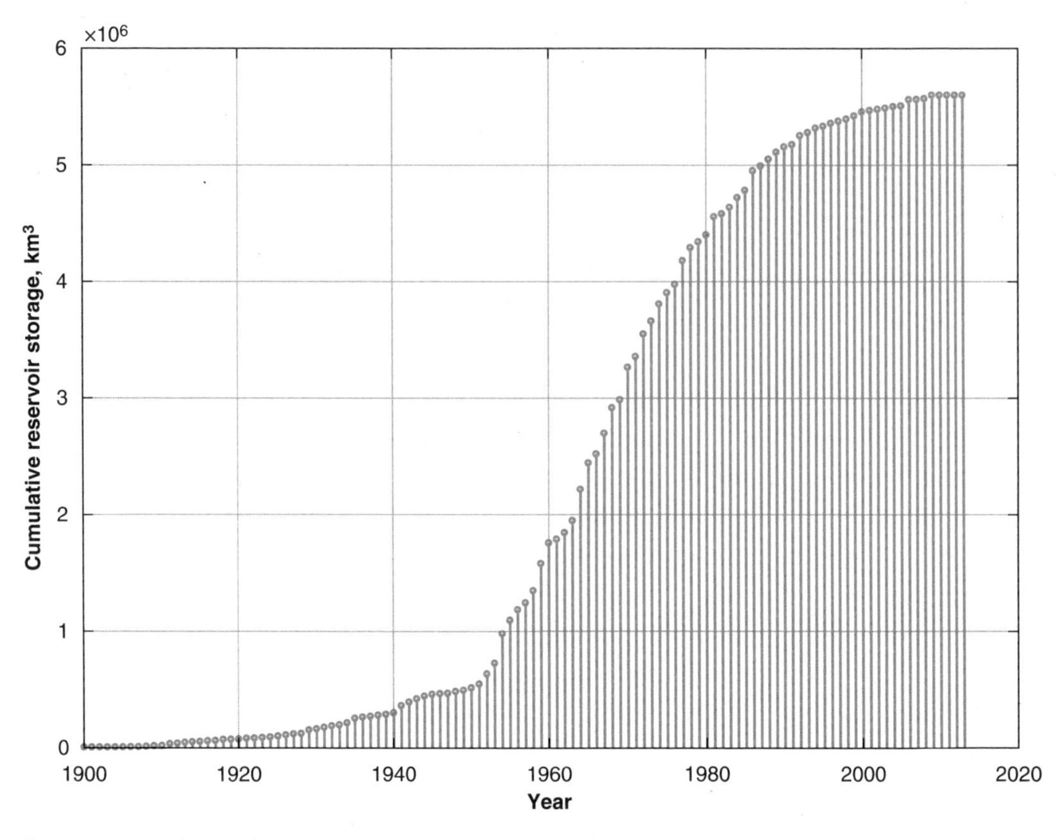

Figure 2.11 Cumulative volume stored in reservoirs globally. Data from the Food and Agriculture Organization 2015.

ecosystems. In addition to the changes in ecosystem services, many reservoirs are constructed in temperate and tropical regions, and the **evaporation** of water from reservoirs is large. The surface area of reservoirs globally is hundreds of thousands of square kilometers, and the average **evaporation** rate over land is several millimeters per year, so the total evaporative loss from reservoirs globally amounts to hundreds of kilometers cubed every year. In addition, large amounts of water can be used for consumption when withdrawn from reservoirs, particularly water used for irrigation, because the **evapotranspiration** losses represent water that is not returned to the river downstream of the dam. Some rivers now even fail to deliver water to their natural outlet—their water is essentially entirely appropriated for **consumptive use** or transfer out of the basin (Figure 2.12).

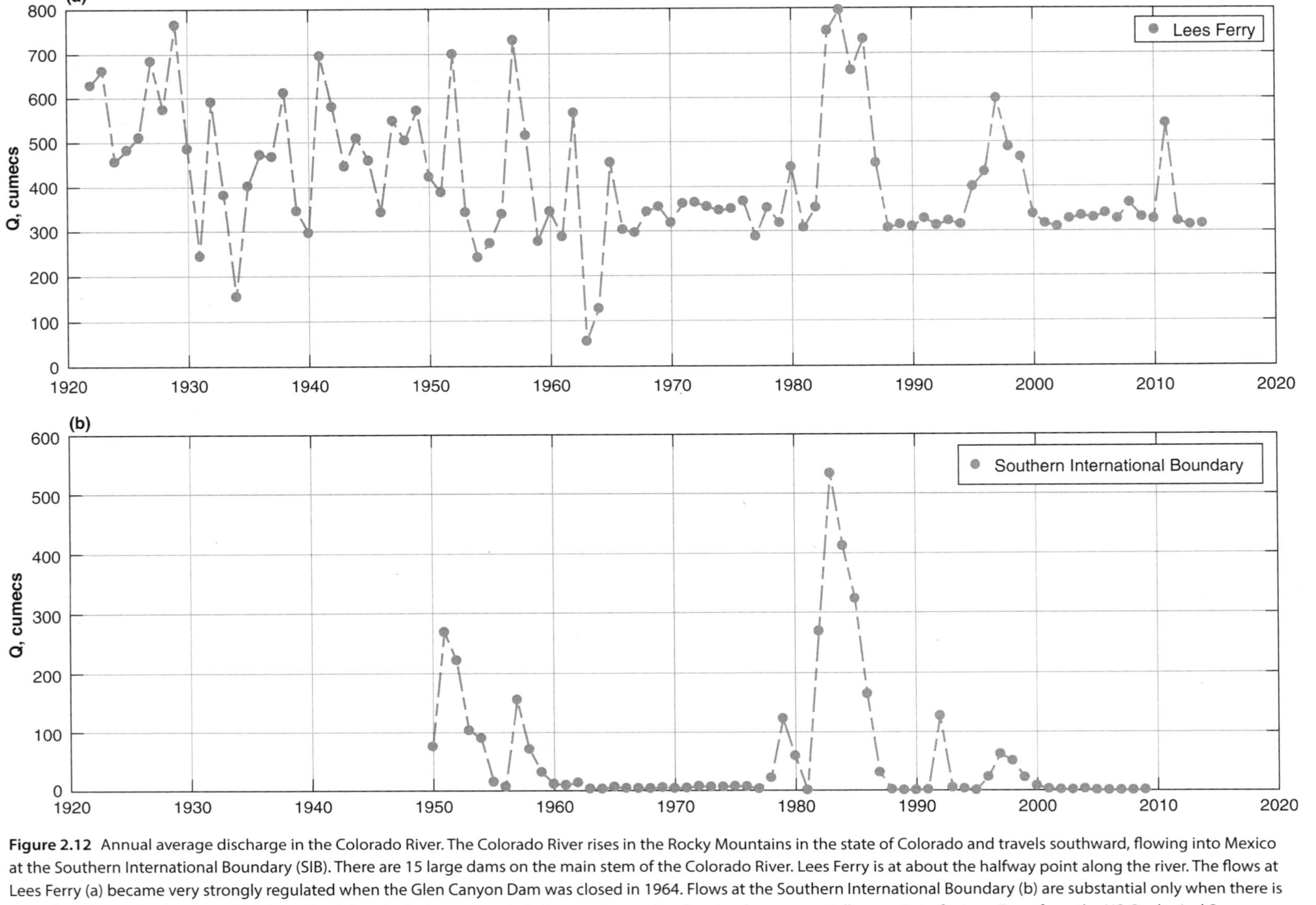

Figure 2.12 Annual average discharge in the Colorado River. The Colorado River rises in the Rocky Mountains in the state of Colorado and travels southward, flowing into Mexico at the Southern International Boundary (SIB). There are 15 large dams on the main stem of the Colorado River. Lees Ferry is at about the halfway point along the river. The flows at Lees Ferry (a) became very strongly regulated when the Glen Canyon Dam was closed in 1964. Flows at the Southern International Boundary (b) are substantial only when there is a surplus of water in the reservoirs along the Colorado during wet periods. In recent years the flow has been essentially zero. Data for Lees Ferry from the US Geological Survey. Data for the SIB are from the International Boundary and Water Commission.

2.6 Concluding Remarks

Surface water from lakes, rivers, and reservoirs is a primary water resource. Fresh water, especially the fresh water at Earth's surface, is a renewable resource, so we get to use it over and over. Nevertheless, humans have appropriated a significant fraction of the resource, and there is an ever-greater need for planning wise water use.

2.7 Key Points

- About 81% of fresh water used by humans is from surface water sources. (Section 2.1)
- There are approximately 18,000 natural lakes in the world with areas exceeding 10 km^2. Because glacial processes formed many of these natural lakes, there is a concentration of them at high latitude. (Section 2.2)
- Wetlands, which provide critical habitat and **ecosystem services**, have declined by over half since 1800, and their destruction or degradation continues to the present. Serious efforts are being made to curtail the damage. (Section 2.2)
- Rivers are the single greatest source of fresh water withdrawn for human use. Measured flows are expressed in cumecs (cubic meters per second) or as an equivalent depth over the drainage basin in a given time. (Section 2.3)
- River flows vary greatly at different places and over time. The likelihood of very high or very low flows occurring—the exceedance probabilities—provides a basis for water resources planning. (Section 2.4)
- The construction of dams and their corresponding reservoirs proliferated around the world in the twentieth century, and many major projects have been completed or are underway in the twenty-first century. (Section 2.5)

2.8 Example Problems

Problem 2.1. Lake Mead is the surface water reservoir above Hoover Dam on the Colorado River. The estimated **discharge** of the river that flows into the reservoir is 585 cumecs. Lake Mead receives about 13 cm of precipitation per year and has a surface area of 640 km^2. Evaporative losses from the Lake are about 65 cumecs. The river outflow from Lake Mead is 375 cumecs. These outflows generate hydroelectricity (via Hoover Dam) and are required for water deliveries to downstream users and for ensuring environmental flows. Lake Mead's storage capacity is 30.5 km^3; increases in storage beyond 30.5 km^3 cannot be stored in the Lake and must be discharged as outflows.

a. Assume Lake Mead is at its full capacity of 30.5 km^3. What is the annual rate of change in storage of Lake Mead based on the aforementioned inflows and outflows?

b. The estimated **discharge** of the river that flows into Lake Mead is likely an overestimate because the estimated discharges are based on a wetter-than-normal period of record. A more likely estimate for annual river inflow to Lake Mead is 352 cumecs. What is the effect on the storage in Lake Mead based on this inflow rate and holding the other aforementioned rates constant?

c. Given the rate of storage change in part *b*, how long would it take until the storage of the reservoir reaches zero?

d. In reality, storage would not become zero. Instead, water would remain in the reservoir because the lowest-elevation intake at Hoover Dam (i.e., the lake outflow) cannot draw water from the reservoir if storage becomes less than 4 km^3. Given this new information, how long would it take until the reservoir reaches a volume of 4 km^3?

e. It is likely that climate change will impact the river discharges into the reservoir. If the river inflow to Lake Mead (estimated to be 352 cumecs) decreases by 20%, what impact would this have on storage?

f. Given the rate of storage change in part *e*, how long would it take until the storage of the reservoir reaches the minimum storage requirements of 4 km^3?

Problem 2.2. Flood frequency analyses typically assume that conditions do not change over time in a statistical sense. That is, the assumption is made that climate has not shifted and made a location much more rainy over time. Even if the assumption that climate has been stable is reasonable for a given location, in an urbanizing area the changes in land use certainly can have a substantial effect on hydrology. Difficult Run flows through an area in Virginia just outside Washington, DC, in the United States. The area was not highly developed in the first half of the twentieth century, but explosive suburban growth took place later. Despite some development of suburban centers in Reston and Vienna, Virginia, the area was still relatively open country before the 1970s. Between 1990 and 2000 the population grew by 19% to nearly 1 million people. This population growth brought with it paved roads, parking lots, and other land changes associated with urbanization. These types of changes have a very pronounced impact on runoff and therefore flood flows. Although urban growth is not instantaneous, and thus there is no absolute cutoff between pre-urbanized and post-urbanized conditions, appreciation

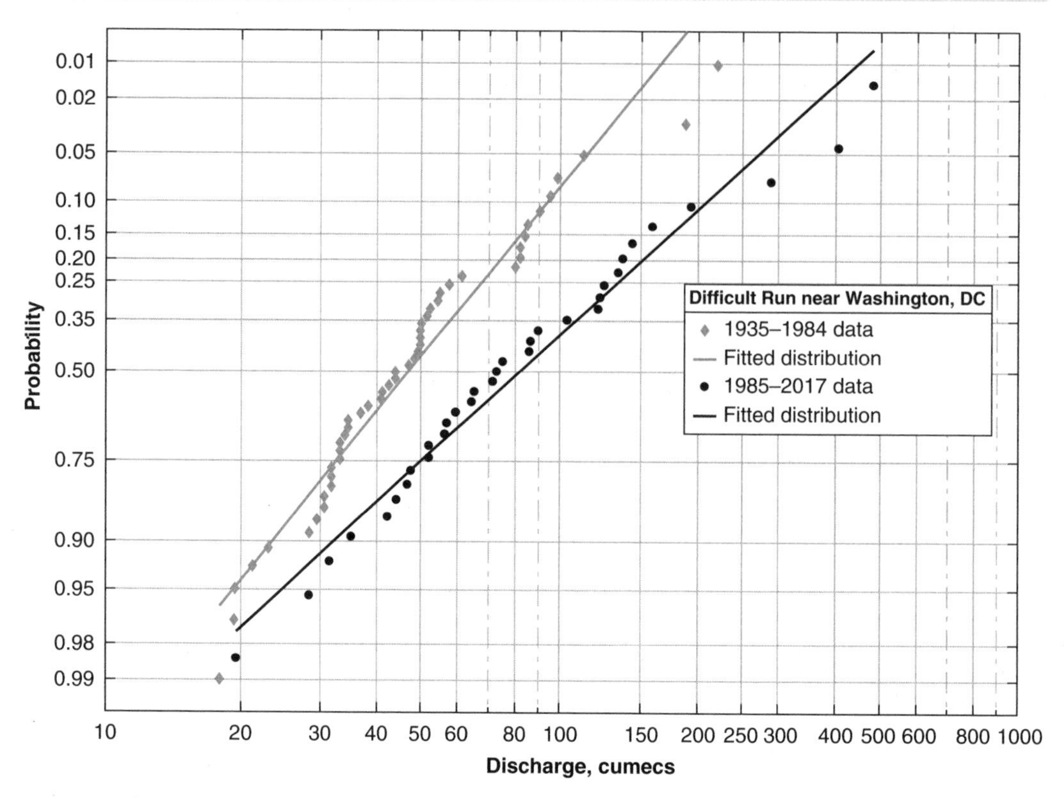

Figure 2.13 Flood frequency curves for Difficult Run near Great Falls, Virginia. Data courtesy of the US Geological Survey. *Note*: The year 1972 was removed from the data set for this analysis; the maximum daily discharge due to Hurricane Agnes was extreme (912 m^3/sec).

of the nature of the changes can be gained by considering the annual floods (maximum **discharge** in any given year) before 1985 and afterward as an indication of changes in flood frequency (Figure 2.13).

a. The 10-year flood has a probability of exceedance of 0.10. From Figure 2.13, estimate the 10-year flood for the period before 1985. Estimate the 10-year flood for the period from 1985 to 2017.

b. The 10-year **discharge** for the period before 1985, $Q10_{1935-1984}$, is expected to be equaled or exceeded once every 10 years on average. Estimate how often the **discharge** $Q10_{1935-1984}$ will be equaled or exceeded in the more recent period—that is, since 1985.

c. If someone bought a house in the 100-year floodplain (**exceedance probability** of 0.01) of Difficult Run in 1930 with a 30-year mortgage, what would have been the estimate for the probability of sustaining flood damage over the full period of the mortgage?

d. What is the same estimate for someone purchasing a house in the pre-1960 100-year floodplain today? (That is, consider the **discharge** for inundating the 100-year floodplain in the recent period to be the same as the **discharge** that defined the 100-year floodplain before 1985. What is the estimated return period for that **discharge** in the post-1984 period?)

Problem 2.3. The city you live in received two 65-year floods within the same year. In 25 words or less, explain why it is possible to get two 65-year floods within 1 year, and include the **exceedance probability** in your answer.

2.9 Suggested Readings

Global HydroLAB. (2014). History of global reservoir and dam construction: 1880 to 2005 [video]. Global Reservoir and Dam (GRanD) database. https://vimeo.com/92062234

Lehner, B., et al. (2011). High-resolution mapping of the world's reservoirs and dams for sustainable river-flow management. *Frontiers in Ecology and the Environment*, *9*(9), 494–502.

2.10 References

Clark, E. A., Sheffield, J., van Vliet, M. T. H., Nijssen, B, & Lettenmaier, D. P. (2015). Continental runoff into the oceans (1950–2008). *Journal of Hydrometeorology*, *16*(4), 1502–1520.

Davidson, N. C. (2014). How much wetland has the world lost? Long-term and recent trends in global wetland area. *Marine and Freshwater Research*, *65*, 934–941.

Downing, J., & Duarte, C. (2010). Abundance and size distribution of lakes, ponds and impoundments. In G. E. Likens (Ed.), *Lake ecosystem ecology: A global perspective* (pp. 469–478). San Diego, CA: Academic Press.

Food and Agriculture Organization (FAO). (2015). AQUASTAT - Georeferenced dams database. http://www.fao.org/nr/water/aquastat/sets/index.stm#dams

Global Runoff Data Centre (GRDC). (2014). *Global freshwater fluxes into the world oceans/Global Runoff Data Centre* (Vol. 1). Koblenz: Federal Institute of Hydrology. http://www.bafg.de/GRDC/EN/03_dtprdcts/31_FWFLX/freshflux_2014_05.html?nn=201790

International Boundary and Water Commission. (n.d.). Colorado River at southerly international boundary in cubic meters per second. https://www.ibwc.gov/wad/DDQSIBCO.HTM

Shiklomanov, I. A., & Rodda, J. C. (Eds.). (2003). *World water resources at the beginning of the 21st century*. Cambridge: Cambridge University Press.

Vorosmarty, C. J., Fekete, B. M., & Tucker, B. A. (1998). Global River Discharge, 1807–1991, version 1.1 (RivDIS). Oak Ridge, TN: Oak Ridge National Laboratory Distributed Active Archive Center. https://doi.org/10.3334/ORNLDAAC/199

CHAPTER 3

Groundwater Resources

3.1 Introduction

Groundwater is by far the largest reservoir of unfrozen freshwater on Earth (see Figure 1.2), but only about one-quarter of the water used by humans comes directly from **groundwater**. Historically, the fraction of **groundwater** used by humans has been much lower. Why is this so? One reason is that **groundwater** is not visible—it is an "out of sight, out of mind" resource. Historically, our understanding of where **groundwater** would occur, how much could be developed, and what the potential consequences would be of development was imperfect at best. A well-known quote from a ruling by a court in 1861 in Ohio in the United States (*Frazier v. Brown*, 12 Ohio St. 294, 1861) about the use of **groundwater** illustrates the point. The "existence, origin, movement, and course of such waters, and the causes which govern and direct their movements, are so secret, occult and concealed, that an attempt to administer any set of legal rules in respect to them would be involved in hopeless uncertainty, and would be, therefore, practically impossible." Secret, occult, and concealed—no wonder science itself had to evolve to inform the large-scale development of **groundwater** resources.

A more pragmatic reason for the lack of extensive, large-scale **groundwater** use before the last 80 or 90 years is the relative difficulty of using **groundwater** compared with surface water. In most instances, **groundwater** simply is not as easily accessible. **Groundwater** springs exist where water flowing below the ground returns to the surface. Where easy access to such **groundwater** springs is available, humans have used **groundwater** as extensively as other surface waters. But in the absence of natural springs, **groundwater** use requires the construction of wells or tunnels. Before the evolution of modern technology, wells were dug by hand. The amount of labor required to hand dig a well meant wells were shallow, and as a result only relatively small quantities of water could be withdrawn. An exception to this observation is the development and use of *qanats*, which are tunnels dug to bring water to the surface as essentially constructed "springs." Qanats supply significant quantities of water and have been used for several millennia for irrigation in arid and semiarid regions.

Even after technological developments allowed deeper wells to be drilled, **groundwater** withdrawals remained limited in most cases due to the energy required to pump water from significant depths. Before the twentieth century, windmills were used extensively to pump **groundwater**, but windmills could deliver only restricted quantities of water at intermittent times. Large-scale exploitation of **groundwater** resources began with the invention of turbine pumps in the first half of the twentieth century. This large-scale human use of fresh water that comes from direct **groundwater** pumping is greater than the amount of **groundwater** recharged in many regions; that is, **groundwater** use in many regions across the globe cannot be sustained in the long run because the resource will be depleted.

3.2 A Little Hydrogeology

Groundwater exists in open spaces in rocks and loosely arranged materials like sand, clay, and soil (Figure 3.1). These open spaces can be the voids between grains of materials, cracks or fractures in rocks, or passages that result from dissolution of a rock such as limestone. Not all water below the ground surface is counted as **groundwater**. **Groundwater** is water that is in rocks or unconsolidated materials

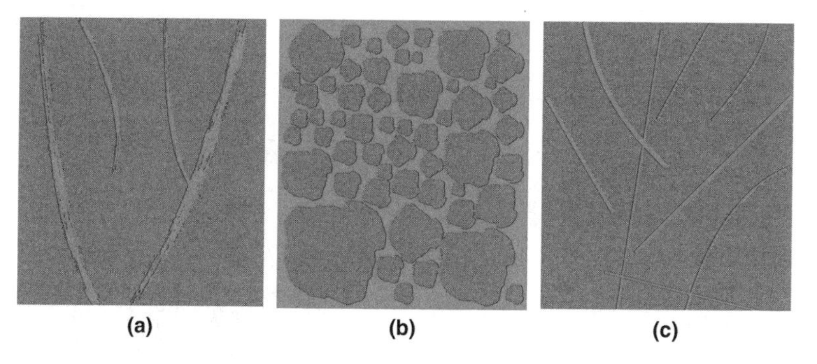

Figure 3.1 Groundwater (blue) fills openings in rocks and soils (gray): (a) solution openings in rocks such as limestone, (b) pores spaces between grains in rocks such as sand and sandstone, and (c) cracks and fractures in rocks such as granite.

saturated with water—in rocks where the openings are completely filled with water.

If you dig a shallow hole in your backyard, it is unlikely to fill with water unless you are adjacent to a river, lake, or ocean. Near the ground surface, the openings in the soil contain some water, but the space is shared with air. In this example, the rock or unconsolidated materials from your dug hole are unsaturated; this water is referred to as soil water, which we will talk about in Chapter 4. If you are industrious and dig a deeper hole, you will likely reach a depth where water does flow into the hole. At this depth, you have reached the **water table**. The **water table** is the dividing line between water-unsaturated and water-saturated rock or unconsolidated materials. **Groundwater** is the water *below* the **water table**; soil water is the water held *above* the **water table**.

Groundwater originates in recharge areas where rain and snowmelt infiltrate into the soil and undergo deep percolation downward through soils and rocks until the water reaches the **water table**. The **groundwater** then flows slowly through geological units known as **aquifers** to discharge points; water always flows from higher **potential** (either higher elevation or higher pressure or both) to lower **potential**. Discharge points can be springs, rivers, lakes, or other surface water bodies (Figure 3.2). **Aquifers** are rocks or unconsolidated materials that can transmit water readily.

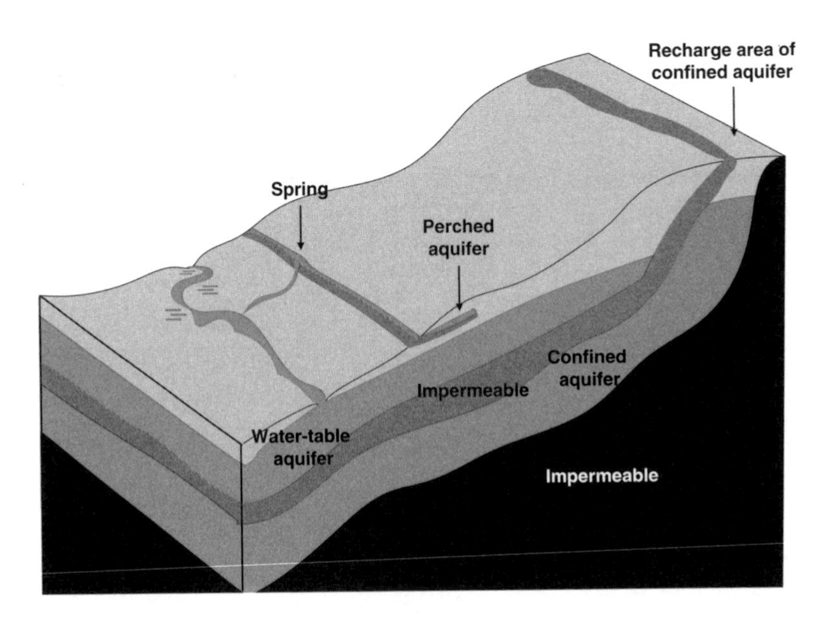

Figure 3.2 Groundwater flows slowly through aquifers from recharge areas to discharge areas.

The amount of water that a rock or other material can hold depends on how many openings there are relative to the amount of solid rock or material. The **porosity** of a rock is the ratio of the volume of open space in the rock to the total volume of the rock. We can also use the measurement of **porosity** for unconsolidated materials. A beach sand might have a **porosity** of 0.3, or 30%. This means that a box of sand 1 meter on a side (i.e., 1 m^3 total volume) would have 0.3 m^3 of void space so it could hold that much water.

Not all rocks are **aquifers**. Some materials such as dense clays or shale can hold significant amounts of water but are essentially impermeable. The void spaces between solid particles are too tiny to allow much water to get through. Other rocks such as unfractured granite are also impermeable, but unfractured rocks do not hold very much water. Impermeable geological units, such as clay and granite, serve as boundaries of **aquifers**. Some **aquifers** have an impermeable bottom as well as an impermeable top. Aquifers with impermeable bottoms and tops are referred to as **confined aquifers**; confined **aquifers** hold water under pressure. For **confined aquifers**, the recharge area may be a considerable distance from the discharge point (see Figure 3.2). **Groundwater** in **confined aquifers** can be thousands of years, or in some cases even hundreds of thousands of years, old.

Aquifers without an impermeable top and **aquifers** that can be recharged by water across their entire extent are referred to as **water-table aquifers** or **unconfined aquifers**. Typically, **water-table aquifers** discharge to springs or streams after following relatively short paths. Some of the water reaches a discharge point such as a stream in a matter of months but more of the water, even in shallow **water-table aquifers**, follows deeper paths and can take decades to reach the discharge area.

3.3 Human Influences on Hydrogeology

Groundwater wells are constructed to improve access to our **groundwater** resources. When a well is constructed to pump water from an **aquifer**, the natural flow system is perturbed. As water is removed from a well, the water level drops, both in the well itself and in the vicinity around the well. This decline in water level is known as **drawdown**, and the depression in the water level is known as the **cone of depression** (Figure 3.3).

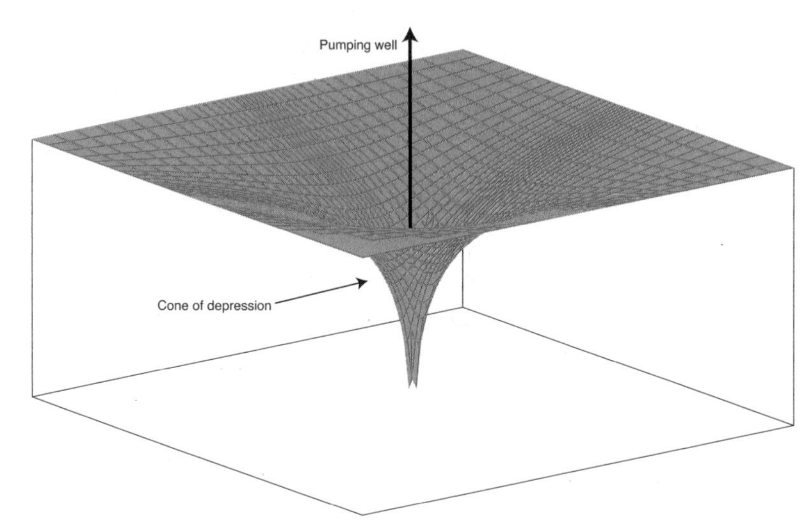

Figure 3.3 The drawdown in water levels in the vicinity of a pumping well creates a cone of depression.

The size of the **cone of depression** is related to the volume of water removed by the well (Box 3.1). In a **water-table aquifer**, the **cone of depression** represents a lowering of the **water table** as the **aquifer** material is drained of water. The region above the **cone of depression** was saturated, but this region now holds air as well as water—as water is removed, air flows in. Only a fraction of the water drains out as the **water table** is lowered, and this fraction is referred to as the **drainable porosity** or the **specific yield**. The volume of the **cone of depression** is equal to the volume of water removed by pumping divided by the **drainable porosity**.

$$V_{cone} = [(\text{Pumping rate}) \times (\text{Time pumped})]/(\text{Drainable porosity})$$

The **drainable porosity** is always less than the total **porosity**. For a sand **aquifer** the **drainable porosity** can be about 25%. If a well is pumped at 100 cubic meters per hour for 10 hours in such an **aquifer**, the expected volume of the **cone of depression** would be (100 m^3/hour × 10 hours)/0.25 = 4,000 m^3.

The water level in a well penetrating a **confined aquifer** is above the top of the **aquifer** and represents the water pressure in the **aquifer**. The level to which water rises in a well is referred to as the **potentiometric surface** (Figure 3.4). When a well is pumped from a **confined aquifer**, the **drawdown** represents the lowering of pressure in the **aquifer**. The **aquifer** itself is not dewatered. So where does the water pumped for the well come from? The recharge area can be a long distance from the well, so water cannot flow quickly

Box 3.1 Groundwater Drawdown Near Dhaka

The city of Dhaka is located in Bangladesh near the eastern end of the Indo-Gangetic Plain (see Figure 3.10). Dhaka derives its municipal water supply from the Dupa Tila aquifer that underlies the city. The aquifer sands are overlain by a thick clay. Recharge through these clay beds is relatively slow. The city underwent tremendous growth from the second half of the twentieth century onward, leading to a similar growth in the abstraction of groundwater (Figure B3.1.1a).

As a consequence of the intense pumping of groundwater, a large cone of depression developed in the aquifer reaching declines of nearly 50 meters by 2002 (Figure B3.1.1b). Groundwater extraction has continued to increase to the present day, and the cone of depression has continued to deepen and expand laterally. The water table beneath much of the city is now well below the clay layer that partially confines the aquifer, so the aquifer is being dewatered—that is, parts of the aquifer are now unconfined. Between 1999 and 2002, the dewatered volume of aquifer estimated from water level measurements in wells was 2,270 million m^3 and the total abstraction of groundwater was estimated to be 2,170 million m^3 (Hoque, Hoque, & Ahmed, 2007). Recall that the specific yield, or drainable porosity, of an aquifer is equal to the ratio of the volume pumped to the volume dewatered. That is, the specific yield is equal to the volume pumped divided by the volume of the cone of depression. Reasonable values for the specific yield of the Dupa Tila sands might be 0.1, 0.15, or perhaps even 0.2. We can use these values to estimate how much of the water pumped in Dhaka from 1999 to 2002 came from decreases in storage; the remaining volume would be an estimate of the volume that came from recharge or capture from the rivers.

For example, if the specific yield were 0.2, we have $0.2 \times 2{,}170 = 434$ million m^3 that resulted in the observed cone of depression. Because the total of groundwater abstractions was 2,270 million m^3, 434 million m^3/2,270 million $m^3 = 0.19$, so with the assumption about specific yield, about 19% of the water came from decreased storage in the aquifer and the remainder came from recharge or capture. In the case of Dhaka, one concern is that the decline in the groundwater table might induce recharge from surrounding rivers, which may have high levels of contaminants.

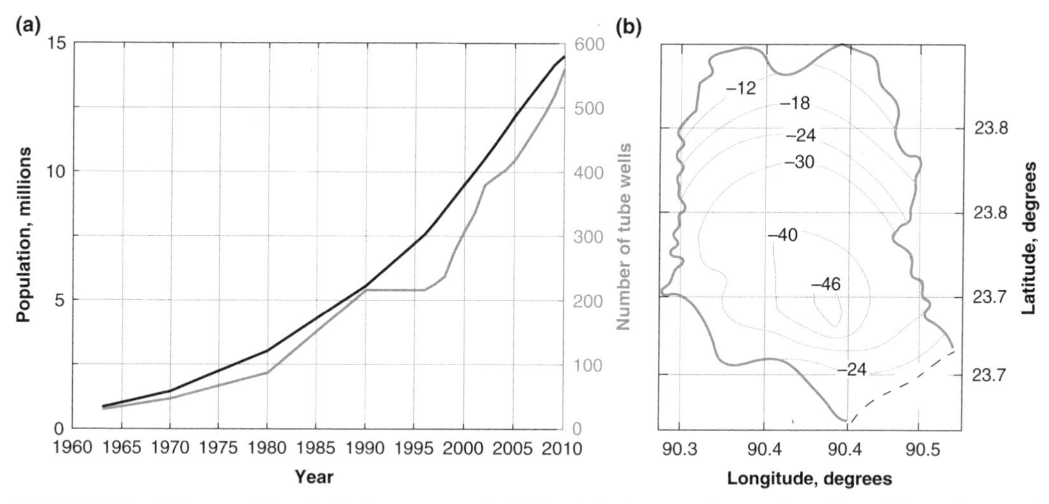

Figure B3.1.1a As the population in Dhaka—a mega-city in Bangladesh—grew, the number of deep tube wells expanded to maintain the water supply. Data from Dhaka Water and Sewerage Authority 2013.

Figure B3.1.1b Drawdown in meters in the aquifer underlying the city of Dhaka in 2002. Dhaka is bordered by rivers except in the southwest. Pumping wells are located throughout the area but with a concentration near the deepest part of the cone of depression. In 2002, the cone of depression was tens of meters deep throughout much of Dhaka. After Hoque et al. 2007.

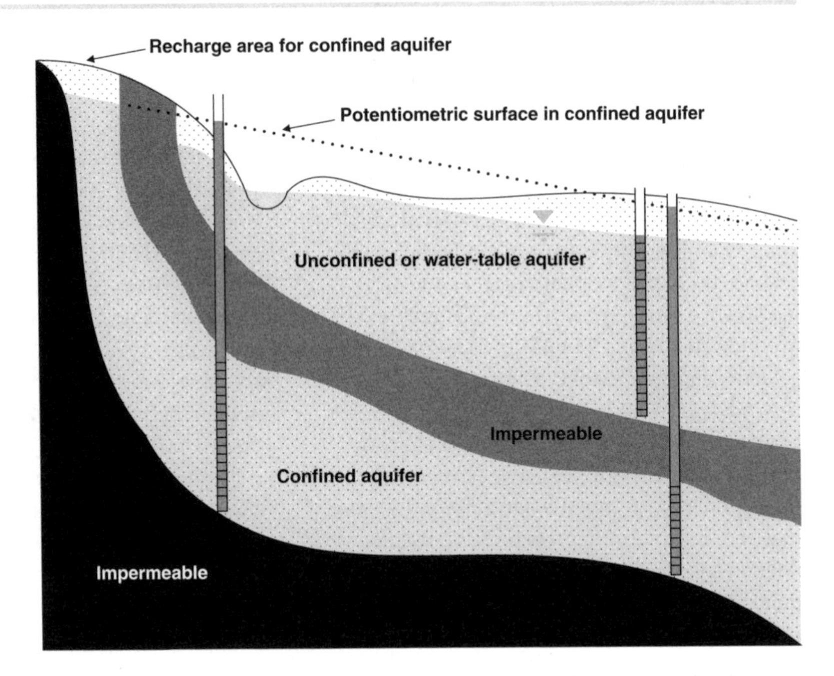

Figure 3.4 Of the three wells shown, the left and far right wells are screened in the confined aquifer, while the center well is screened in the unconfined aquifer. In confined aquifers, the water level in the wells (left and far right) rises above the top of the aquifer and indicates the position of the potentiometric surface. In unconfined aquifers, the water level in the well (center) indicates the height of the water table, which is also the position of the potentiometric surface for the unconfined aquifer. Redrawn from Hornberger et al. 2014.

enough from the recharge area to the well. The answer is that the water removed by pumping is due to the expansion of water as pressure is reduced. Water is not totally incompressible, so it does expand slightly when pressure is released. Because the top of the **aquifer** is partly supported by the water pressure, the **aquifer** has to bear more of the weight of the overlying materials. As a result of this weight, materials are compressed slightly, because, like water, these materials are also not totally incompressible and this slight compression forces water to flow to the pumping well.

In a **confined aquifer**, the volume of the **cone of depression** is not related to the **drainable porosity**, because the **aquifer** is not drained. **Drawdown**, the lowering of the water level in wells, is related to the **storativity** of the **confined aquifer**. **Storativity** is the volume of water released per volume increment in the **cone of depression**; **storativity** is a dimensionless quantity. The **storativity** depends on the compressibility of water and rock. Compressibility values for water and rocks are small numbers, so the **storativity** of **confined aquifers** is very small, typically in the range of 10^{-3} to 10^{-5}. The volume of the **cone of depression** in a **confined aquifer** is equal to the volume of water removed by pumping divided by the **storativity**.

$$V_{cone} = [(\text{Pumping rate}) \times (\text{Time pumped})]/(\text{Storativity})$$

For a **confined aquifer** with **storativity** 10^{-5}, our example calculation for pumping at 100 cubic meters per hour for 10 hours results in a **cone of depression** with volume (100 m^3/hour × 10 hours)/0.00001 = 100,000,000 m^3. Pumping from confined aquifers can result in wide and deep cones of depression.

3.4 Groundwater Resources

As noted in Chapter 1, **groundwater** represents the largest fraction of fresh water on Earth except for icecaps and glaciers. How do we know how much liquid water is stored beneath the surface? We have to infer the total amount often using crude estimates. For example, the total continental land area is about 148×10^6 km^2. If we guess that the total depth of porous rock is about 1 km and that the average **porosity** of the rock is 0.15, we derive an estimate of the total volume of **groundwater** stored as about 22 million km^3. A somewhat more sophisticated version of this calculation gives 23.4 million km^3 (Shiklomanov & Rodda, 2003). This estimate includes all **groundwater**—saline as well as fresh. The total volume of fresh **groundwater** is about 45% of the estimated total or about 10.5 million km^3 (see Table 1.1).

Of course, **groundwater** is not uniformly distributed across continents. A large portion of Earth's **groundwater** is contained within major **aquifer** systems. Refined estimates of **groundwater** storage require information on specific **aquifer** systems. **Aquifer**-systems analyses indicate that the previous estimates using crude assumptions likely result in overestimates (Richey et al., 2015). Nevertheless, there are large volumes of fresh **groundwater** globally.

Because **groundwater** is a complex resource, it is often difficult to fully apportion resources as renewable or nonrenewable. In some aquifers around the world, more water is removed than recharged, suggesting that **groundwater** is being overexploited. If water that is removed by pumping can be naturally replenished on a time scale of months, years, or decades, the resource may be considered renewable for the purposes of water planning and sustainable management. The concept of **groundwater** as a renewable resource must include the link between **groundwater** and surface water because lowering the **water table** can have serious consequences for the flow in streams and rivers—consequences that include impacts on ecosystems and on surface water withdrawals for multiple purposes. If the times for replenishment of **groundwater** are hundreds or thousands of years, the resource is essentially nonrenewable for purposes of water planning and management. Extraction of water from aquifers with "nonrenewable" **groundwater** resources is considered **groundwater mining**.

The rate of renewal of **groundwater**, expressed as a volume per year, is the average amount of water that is available for net recharge in a year: the water that precipitates from the atmosphere minus the amount that is returned to the atmosphere by evapotranspiration and minus the amount that runs off in rivers. Estimates of the rates of renewal of **groundwater** are based on calculations of the overall water budget, similar to that described in Section 1.6. Estimates of the renewal rates for large countries that cover a variety of climate zones can exceed 1,000 km^3 per year whereas the amount is quite small for smaller arid countries (Figure 3.5).

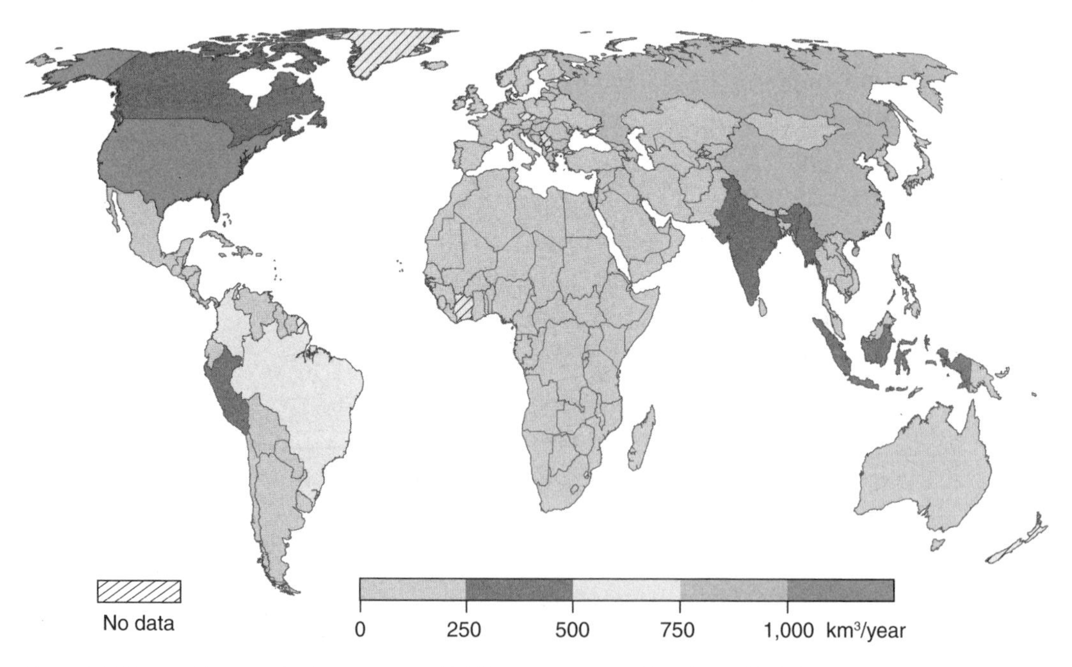

Figure 3.5 Renewable groundwater reserves, cubic kilometers per year. Data from FAO 2016.

3.5 Groundwater Use

Groundwater has many beneficial uses across the globe. It provides a safe, clean supply of water for many villages, towns, and municipalities. **Groundwater** serves as a source of irrigation water for crops in semiarid areas, enabling the production of local food supplies. It is a reliable source of water for industrial purposes, such as providing cooling water for thermoelectric power plants. The growth in human population, the increase in urbanization of the world, and the expansion of agriculture all have resulted in increased extraction of **groundwater** (Figure 3.6). **Groundwater** wells can be constructed essentially anywhere above an **aquifer**, making it ideal for use at a desired point without the need for long pipelines or other infrastructure. One major reason for the growth of **groundwater** use is that modern technologies for constructing wells and pumping water have made exploitation of reserves relatively cheap.

The total **groundwater** withdrawals by countries as of 2015 ranged from negligible to 250 km³ per year (Figure 3.7). On average

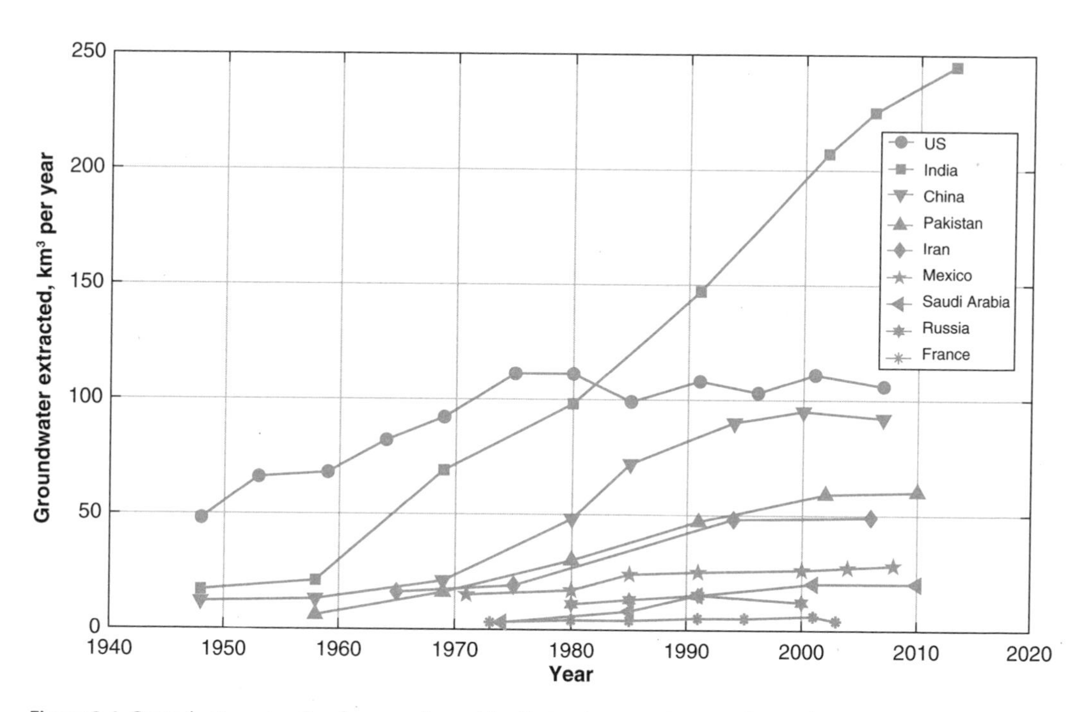

Figure 3.6 Groundwater extraction for several countries that make extensive use of groundwater resources. Data from van der Gun 2012.

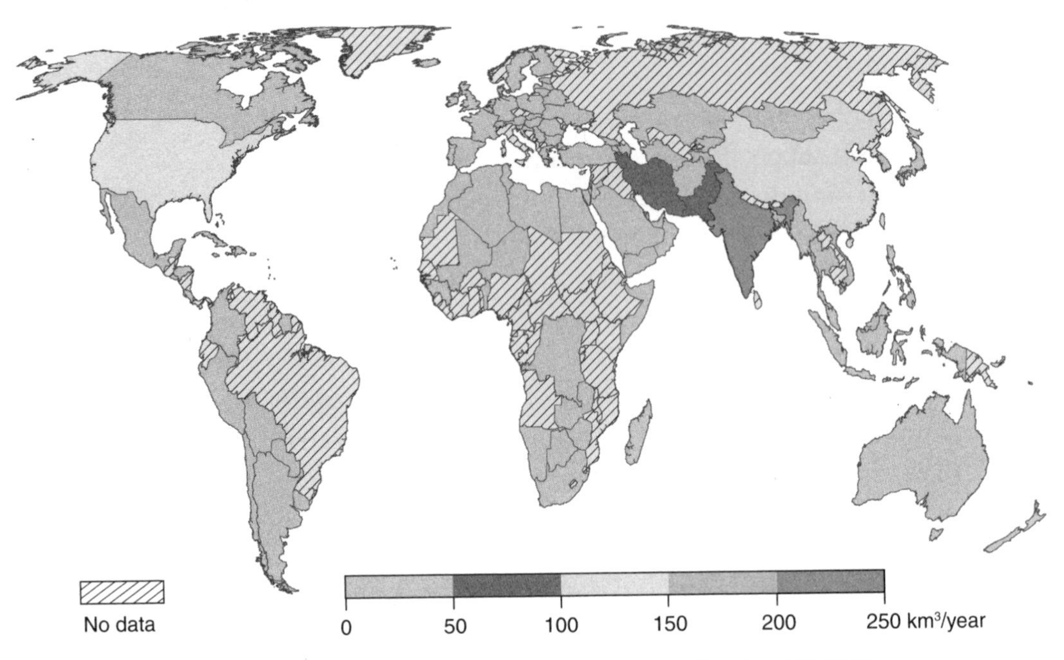

Figure 3.7 Fresh groundwater withdrawals in 2015, cubic kilometers per year. Data from FAO 2016.

about 71% of global **groundwater** withdrawals are for agriculture (irrigation), and the remaining 29% are for municipal and industrial uses. For the top six countries in terms of total **groundwater** withdrawal, the percentages of groundwater used for agriculture are 89% (India), 71% (United States), 54% (China), 90% (Pakistan), 60% (Iran), and 64% (Mexico).

Aquifers can be used to store water during times when surface water is plentiful, so it can be retrieved when needed. That is, **groundwater** recharge can be artificially enhanced by spreading the water into **infiltration** basins or by injecting water through recharge wells (the reverse of pumping). This practice is referred to as **managed aquifer recharge** and has been used successfully for over a century in many regions of the world (Casanova et al., 2016).

3.6 Groundwater–Surface Water Interactions

In considering **groundwater** resources, it is important to keep in mind that there is not a sharp division between surface water and **groundwater**. Streams, lakes, and wetlands all are **groundwater** discharge areas. Thus, if **groundwater** that normally flows into a surface water body is intercepted by pumping before it discharges, the amount of **groundwater** that discharges into surface water bodies

Box 3.2 Groundwater Pumping and Streamflow

Spring Creek in southwest Georgia in the United States is at the northern end of the Floridan aquifer system (see Figure 3.10). The summer low flows in Spring Creek are discharges from the Upper Floridan aquifer. Starting in the 1970s, irrigation, fueled by pumping from the Floridan aquifer, expanded tremendously across the Spring Creek catchment. The rate of groundwater withdrawals for irrigation has decreased flows into the river. In fact, the groundwater withdrawals have actually resulted in reducing the flows to zero during some periods—the groundwater withdrawals have resulted in drying the stream completely (Figure B3.2.1).

Between 1938 and 1980, the records for Spring Creek at Iron City showed no days with zero flow. From 1980 to 2014, the records showed 424 days with no flow—zero cumecs. This depletion of streamflow has led to ecological disruptions. Although a moratorium on new permits for groundwater pumping were put in place in 2014, the existing permits remain in effect. Farmers have invested heavily in irrigation infrastructure, and they depend on irrigation to provide high crop yields, making reversal of the flow reductions in Spring Creek a difficult issue.

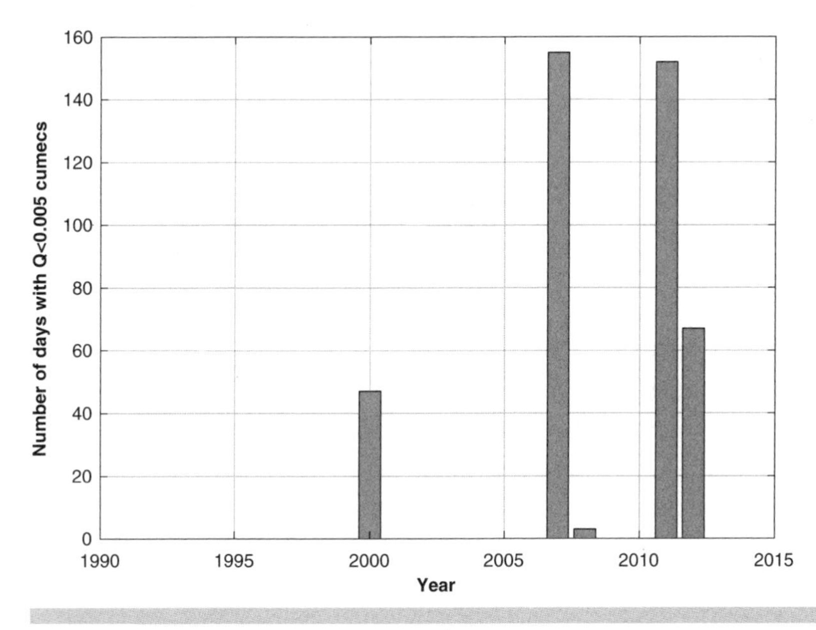

Figure B3.2.1 Days with discharge below 0.005 cumecs recorded in Spring Creek at Iron City. Note that there were no days with flow this low before 1980. Data from US Geological Survey.

will be impacted. This can result in a decline in surface water levels, impacting ecosystems and riparian zones as well as surface water availability for human uses (Box 3.2). In fact, if the well is close enough to a stream, the **drawdown** can extend to the stream; instead of having **groundwater** flow to the stream, stream water can be induced to recharge the **aquifer** (Figure 3.8). We refer to a river that recharges **groundwater**, either because of natural gradients that cause flow from the river to the **aquifer** or flow induced by pumping, as a losing river.

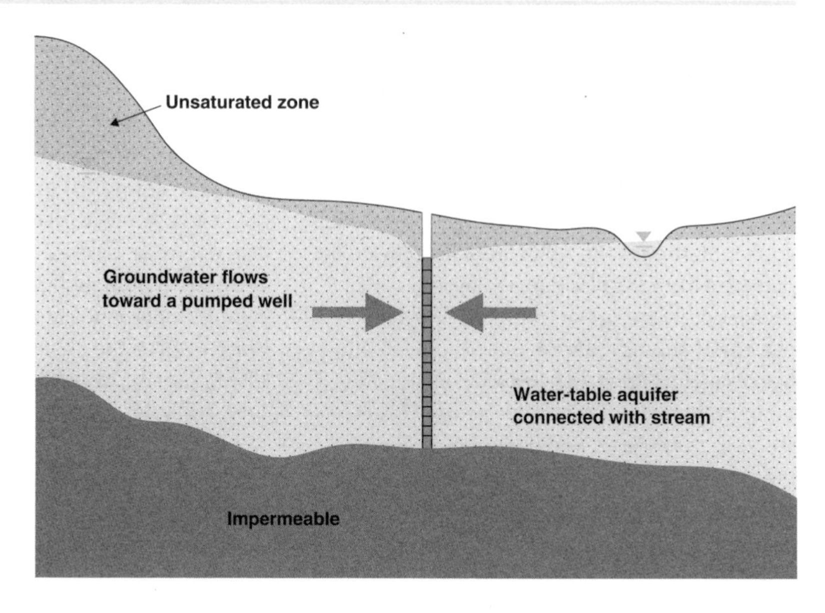

Figure 3.8 Drawdown at a pumped well can cause flow from a nearby stream into the aquifer.

3.7 Groundwater Depletion and Safe Yield

It is unquestionable that water that is removed from aquifers lessens the amount of **groundwater** stored in the formation. The decline in storage is **groundwater depletion**. The **safe yield** of an **aquifer** is defined as the amount of depletion that can be assigned without incurring unacceptable consequences or undesirable results. Note that safe yield is not equal to recharge, one reason being that maintaining natural groundwater discharges is essential for preserving important ecosystem services (Chapter 8). Excessive **groundwater depletion** can have many adverse consequences. As a result, there are many undesirable results that stem from **groundwater depletion** that must be considered in estimating **safe yield**.

Groundwater depletion lowers water levels and thus increases costs for pumping water—an economic consequence that can be very important. Depletion can induce recharge from saline or contaminated sources of water—a water quality consequence. Lowering of **groundwater** levels through depletion can reduce flows to streams and wetlands and even drain them—an ecological consequence. When **groundwater** use affects flows in rivers, there may also be consequences for the water users who rely on surface water availability. Declining **groundwater** levels can also cause wells to go dry; when **groundwater** levels decline, the well may no longer be deep enough to access **groundwater** resources. Water withdrawn

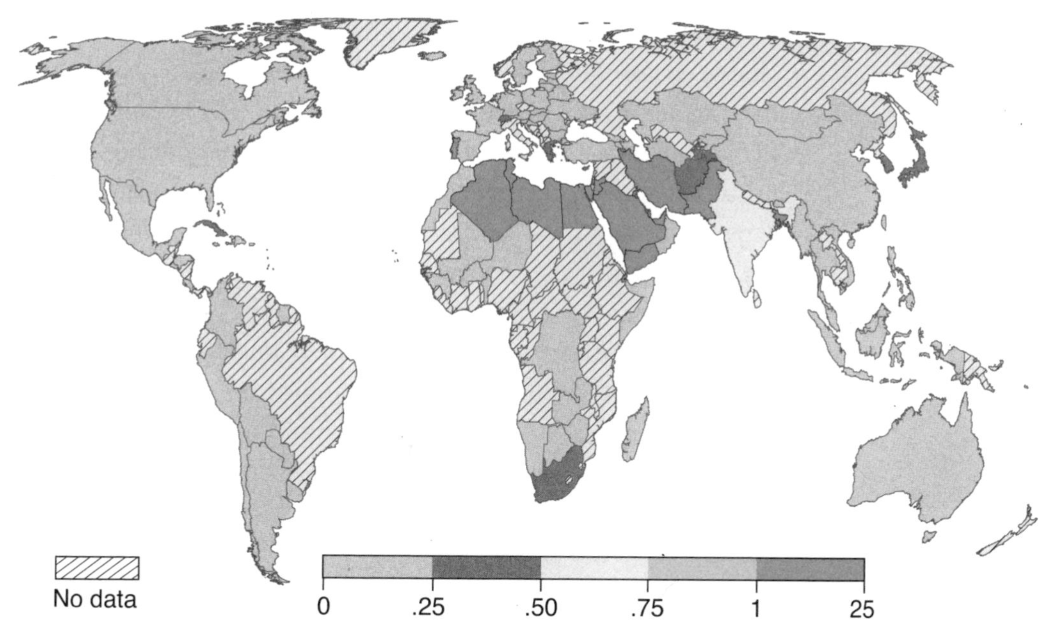

Figure 3.9 Ratio of 2015 groundwater withdrawals to renewable groundwater. The highest ratios are for countries for which withdrawals exceed recharge; for example, United Arab Emirates (ratio of 23), Kuwait (ratio of 21), Saudi Arabia (ratio of 10), and Libya (ratio of 9). Data from FAO 2016.

from **confined aquifers** leads to changes in water pressure that can dewater clay layers in the system; the compaction of these layers results in land subsidence (the lowering of the land surface), which can have negative impacts on infrastructure.

Excessive **groundwater depletion** is occurring throughout the world. One index that points to the issue is the ratio of **groundwater** withdrawals (see Figure 3.7) to renewable **groundwater** (see Figure 3.5). In extreme cases this ratio will exceed unity; that is, more water is being removed than is being recharged (Figure 3.9). For most of the regions of the world where the withdrawals are greater than the recharge, water is being removed from semi-confined or confined aquifers that contain fossil water. When **groundwater** is mined, the issues surrounding depletion involve the best use of the resource for the current generation as well as for future generations.

For many countries, the ratio of withdrawal to renewable resource is less than 1 (Figure 3.9). For example, the country averages for Japan, India, Belgium, and Israel are 0.35, 0.58, 0.69, and 0.80, respectively. A value of the ratio less than 1 does not mean that excessive depletion is not occurring. First, the **safe yield** of any

aquifer is not the recharge rate to that aquifer. **Groundwater** discharge occurs where **groundwater** flows into streams, rivers, and the ocean. In places where **groundwater** and surface water are connected, excessive pumping can decrease **groundwater** discharge to surface waters, likely resulting in undesirable consequences to ecosystems and river flows. Second, countrywide averages will mask any local or regional variation. That is, one region of a country may have very little extraction of **groundwater** while another may experience excessive depletion. For example, the countrywide average rate of withdrawal in the United States is only 8% of the renewable resource, but in several regional aquifers the value exceeds 100%.

There are many aquifers of regional and local importance, several of which are delineated in Figure 3.10. The recent rates of depletion for a selected set of the delineated aquifers indicate that current **groundwater** use is not sustainable over the long run in many places, such as the Central Valley (3.7 km^3 per year), the Great Artesian Basin (0.3 km^3 per year), the High Plains Aquifer (12.4 km^3 per year), the North China Plain (6.1 km^3 per year), the central swath of the Indo-Gangetic Plain (54 km^3 per year), the North Sahara (1.5 km^3 per year), and the Nubian (2.4 km^3 per year) (van der Gun, 2012). In some areas, such as the North China Plain (Box 3.3),

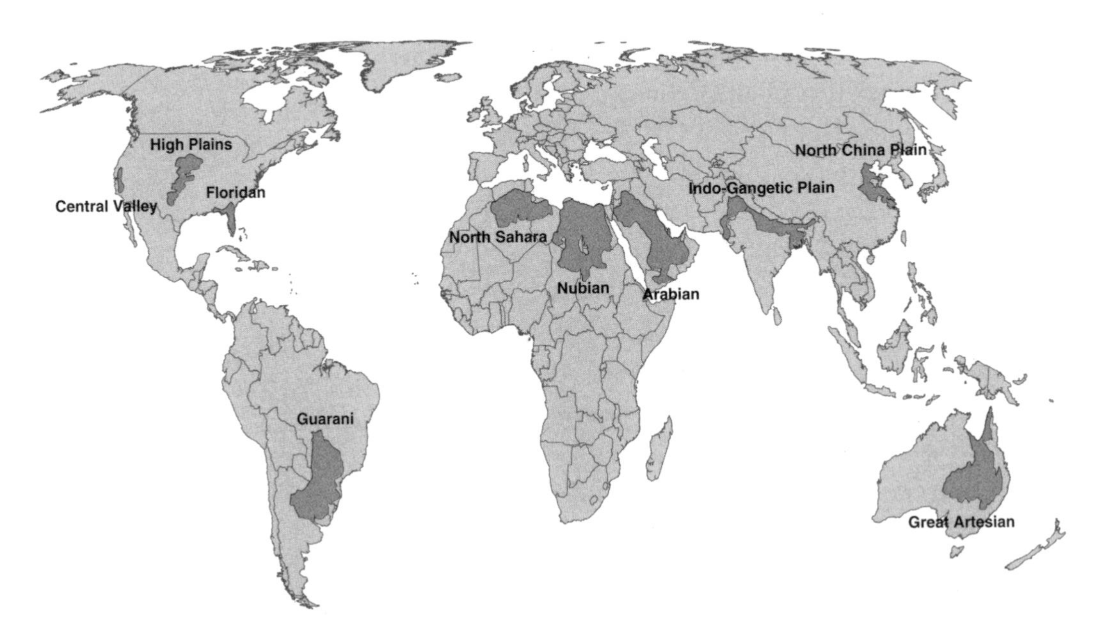

Figure 3.10 Selected aquifer systems with large quantities of groundwater that can be extracted.

Box 3.3 Groundwater Depletion in the North China Plain

The North China Plain (NCP) is the largest grain producing region of the People's Republic of China (see Figure 3.10). Grain yields increased steadily in the NCP from the mid-twentieth century to the present as irrigation using groundwater increased and fertilizer application increased by several hundred percent (Pei et al., 2015). Groundwater depletion occurs throughout the region, with average declines in the water table of about one-third of a meter per year on average (Figure B3.3.1). With the largest population in the world, the People's Republic of China must sustain the agricultural production in the NCP, but to do so, the country must determine how to sustain the agricultural production while considering and addressing groundwater quantity and quality concerns.

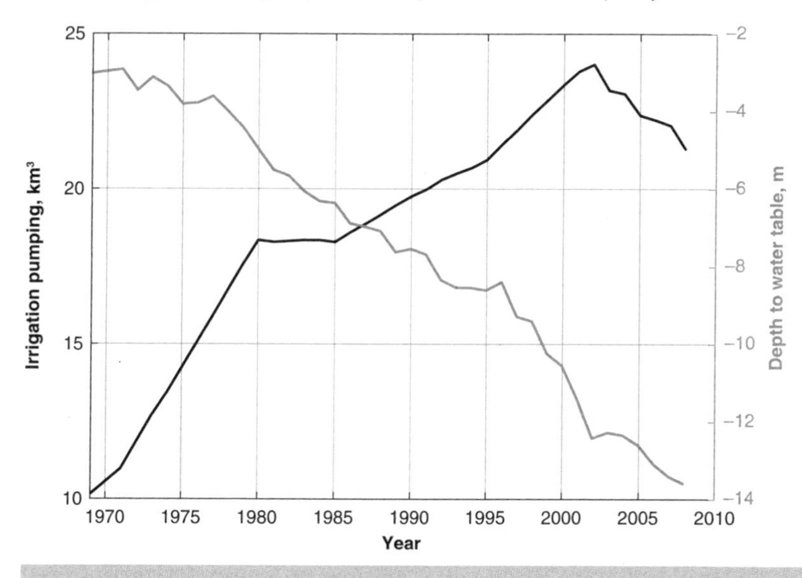

Figure B3.3.1 Irrigation pressure in the North China Plain increased by more than a factor of two since 1970 with a concomitant average decline in the water table of more than 10 meters. Data from Pei et al. 2015.

an agreement on a value of **safe yield** followed by limitations on **groundwater** withdrawals for irrigation could lead to cessation of depletion. In some areas such as northern Africa (the Nubian Sandstone aquifer), the **groundwater** is essentially all very old, so any use is **groundwater mining**. The reserves of the Nubian Sandstone aquifer are very substantial, so wise use can continue for a long time, perhaps hundreds or thousands of years. This does not mean that adverse impacts will not occur locally. Oases—areas of **groundwater** discharge—can decline or even cease to exist as discharge zones. The Guarani Aquifer in South America has not been heavily impacted to date, but there is a high level of concern about its management given rapid development and population growth in the area. In 2010, an international agreement was signed by Argentina, Brazil, Paraguay, and Uruguay that is aimed to be a framework for governance of this large **aquifer** (Green, 2012).

3.8 Concluding Remarks

Fresh water in aquifers is a large fraction of the total available resource. A relatively small portion of this water is renewable for planning and management purposes, which still represents a very substantial volume across the globe. Until the middle of the twentieth century, **groundwater** use was generally at a sustainable level. As well drilling methods improved and powerful pumps became available, large-scale pumping increased, which led to depletion of many aquifers. The issues faced by water resource planners and managers include loss of surface water due to pumping in aquifers that provide flow to streams and wetlands, the intrusion of salt water into aquifers near the coasts because of excessive **groundwater** pumping, and widespread land subsidence. The management of **groundwater** resources within a given governance framework (e.g., a system of laws and regulations—Chapter 11) has become a critical issue across the world.

3.9 Key Points

- Exploitation of **groundwater** resources increased explosively over the last century as technological advances were made in well construction and water pumps. (Section 3.1)
- Aquifers, geological formations that are sources of **groundwater**, are either unconfined (i.e., **water table** aquifers) or confined aquifers. (Section 3.2)
- A **cone of depression** develops around a well that pumps water from an **aquifer**. The amount of water that is removed from an **aquifer** relative to the volume of the **cone of depression** is the **specific yield** or **drainable porosity** of a **water table aquifer** and is the **storativity** of a **confined aquifer**. (Section 3.3)
- The total amount of fresh water stored underground is very large but only a fraction of it is accessible for use. The renewal rate of **groundwater** is determined by the rate at which water is added to the **aquifer** by net recharge. (Section 3.4)
- **Groundwater** is a valuable resource and serves many beneficial purposes. Agriculture uses 70% of all **groundwater** on average, making **groundwater** very important for food security. (Section 3.5)

- **Groundwater depletion** is occurring across the globe, leading to concern about how to manage the resource sustainably. (Section 3.7)
- The **safe yield** of an **aquifer** is *not* equal to the recharge. **Safe yield** intends that adverse impacts be minimized and even totally avoided in cases where they are judged to be critical. To determine **safe yield** many things must be considered, including impact on surface water, induced recharge of saline or contaminated water, and land subsidence. (Section 3.7)

3.10 Example Problems

Problem 3.1. Pumping tests are often done in aquifers to estimate how they will respond to extraction of **groundwater**. The tests are done by pumping water out of one well and observing how water levels change in nearby wells. The measurements can be analyzed to determine the volume of the **cone of depression** at a given time. A pumping test was done near Kearney, Nebraska, in the United States at a time with little or no recharge. Nebraska is in the northern part of the High Plains Aquifer; see Figure 3.10 for the location of the High Plains Aquifer. A well was pumped at a rate of 250 m^3 per hour for 24 hours. The volume of the **cone of depression** was determined to be 58,500 m^3 at the end of this time.

a. What is the **specific yield** of this **water-table aquifer**?

b. What would the volume of the **cone of depression** have been after 10 hours of pumping?

Problem 3.2. The volume of water stored in an **aquifer** divided by the net rate of depletion gives a time to total **aquifer** depletion if nothing changed. But as an **aquifer** is depleted, it becomes more difficult and more expensive to extract water. Thus, it is not reasonable to assume that 100% of **aquifer** storage would actually be extracted. Nevertheless, knowledge about a time frame can focus attention on regional **groundwater depletion** issues. The depletion rate in the North Sahara Aquifer is estimated to be about 1.5 km^3 per year (Section 3.7). The areal extent of the **aquifer** is estimated to be 1,280,000 km^2. The average thickness of the **aquifer** is not known with precision, but it is probably between 500 and 1,000 m. The **specific yield** of the **unconfined aquifer** may be between 0.05 and 0.1.

a. If the current rate of abstraction proceeds into the future, estimate how long would it take to deplete 25% of the accessible water in the **aquifer**.

b. Repeat part *a* with the assumption that the net withdrawal increases by a factor of 10.

Problem 3.3. The North China Plain is the largest grain-producing region of the People's Republic of China. Grain yields increased steadily in the NCP from the mid-twentieth century to the present as irrigation using **groundwater** increased and fertilizer application increased by several hundred percent (Pei et al., 2015). **Groundwater depletion** occurs throughout the region, with average declines in the **water table** of about one-third of a meter per year on average (see Box 3.3). With the largest population in the world, the People's Republic of China must sustain the agricultural production in the NCP while also considering **groundwater** quantity and quality.

The large fraction of the **groundwater depletion** in the North China Plain is due to pumping for irrigation, largely for wheat and maize. Suppose that 6 million hectares are irrigated with a net consumption of 100 mm/year.

a. How much of the irrigation would have to be curtailed to reduce the estimated rate of **groundwater depletion** to 4.15 km^3/year?

b. The NCP is very important as a source of grain. If the grain yield is 19 kg/hectare per unit of water used, what is the total loss of grain?

c. If the average grain share needed is 400 kg/person, for how many people would an alternate source of grain be needed?

3.11 Suggested Reading

Konikow, L. F., & Kendy, E. (2005). Groundwater depletion: A global problem. *Hydrogeology Journal, 13*(1), 317–320.

3.12 References

Casanova, J., Devau, N., & Pettenati, M. (2016). Managed aquifer recharge: An overview of issues and options. In A. J. Jakeman, O. Barreteau, R. J. Hunt, J. D. Rinaudo, & A. Ross (Eds.), *Integrated groundwater management.* Cham, Switzerland: Springer International Publishing.

Dhaka Water and Sewerage Authority. (2013). *Annual Report 2012–2013.* Dhaka. Retrieved from http://dwasa.org.bd/annual-reports/

Food and Agriculture Organization (FAO). (2016). AQUASTAT Main Database, Food and Agriculture Organization of the United Nations (FAO).

Green, B. A. (2012). Advances in transboundary groundwater management: The Guarani Aquifer Agreement. *New England Law Review, 46,* 61–69.

Hoque, M. A., Hoque, M. M., & Ahmed, K. M. (2007). Declining groundwater level and aquifer dewatering in Dhaka metropolitan area, Bangladesh: Causes and quantification. *Hydrogeology Journal, 15*(8), 1523–1534.

Hornberger, G. M., Wiberg, P. L., Raffensperger, J. P., & D'Odorico, P. (2014). *Elements of physical hydrology* (2nd ed.). Baltimore: Johns Hopkins University Press.

Pei, H., Scanlon, B. R., Shen, Y., Reedy, R. C., Long, D., & Liu, C. (2015). Impacts of varying agricultural intensification on crop yield and groundwater resources: Comparison of the North China Plain and US High Plains. *Environmental Research Letters, 10*(4), 44013.

Richey, A. S., Thomas, B. F., Lo, M.-H., Famiglietti, J. S., Swenson, S., & Rodell, M. (2015). Uncertainty in global groundwater storage estimates in a total groundwater stress framework. *Water Resources Research, 51*, 5198–5216.

Shiklomanov, I. A., & Rodda, J. C. (Eds.). (2003). *World water resources at the beginning of the 21st century.* Cambridge: Cambridge University Press.

van der Gun, J. (2012). *Groundwater and global change: Trends, opportunities and challenges.* Paris: UNESCO.

CHAPTER 4

Soil Water Resources

4.1 Introduction

In considering water as a resource for humans, soil water is different than surface water or **groundwater** in the sense that we do not extract water directly as a liquid from soils. Rather, soil water is extracted by plants. It is this fundamental underpinning of the entire food chain that makes soil water resources so essential to humans. What exactly is soil water? Soil water is the water between the soil surface and the **groundwater** table. Soil water, like **groundwater**, is held in openings in the soils and rocks, but these openings are partly filled with water and partly filled with air, unlike the openings below the **groundwater** table. Water infiltrates into this **unsaturated zone**, also referred to as the vadose zone, at the surface of Earth. Some of the infiltrated water migrates downward to the water table as **groundwater** recharge, but a portion of the water is retained in the soils and rocks in the **unsaturated zone** by capillary forces. This is similar to your kitchen sponge: some water is retained in a kitchen sponge after excess water has drained away into the sink. Thus, soils act both to regulate the flow of water from rainfall and snowmelt to **groundwater** and to store water that is crucial to the functioning of terrestrial ecosystems.

4.2 A Little Soil Science

The amount of water available to infiltrate the soil and the rate at which plants take up water from the soil depend on climate and weather conditions as well as on the nature of the soil. Evaluation of soil water resources necessarily requires a combined understanding of soil properties and climate. In soils, the term *texture* is used to describe how soil pore space varies depending on the proportions of sand, silt, and clay in the soil. Smaller grain size means smaller pores. For example, let us consider the "pores" between

Figure 4.1 Soil texture is defined by fractions of sand, silt, and clay present. Fine sand is about 10 times smaller than coarse sand, silt is about 10 times smaller than fine sand, and clay is about 10 times smaller than silt. Note that loam soil is a mixture of sand, silt, and clay.

bowling balls packed in a crate and the pores between marbles packed in a similar sized crate. The pores between the bowling balls are much larger than the pores between the marbles. The size of different soil particles, and thus their pore sizes, varies across orders of magnitude (Figure 4.1). The diameters of the mineral grains in a coarse sand are between 0.2 and 2.0 millimeters. The comparable diameters for fine sand and silt grains are 0.2 to 0.02 mm and 0.02 to 0.002 mm, respectively. Mineral grains of clay-sized particles are less than 0.002 mm in diameter.

Water is held more tightly in small soil pores relative to large soil pores. In a glass tube, capillary rise is inversely related to the diameter of the tube. The capillary forces holding water in a tube are stronger for small diameter tubes than bigger tubes. This also is an intuitive notion if you are familiar with kitchen sponges: a sponge with large openings will retain much less water after draining than will one with very small openings. The capillary forces exert a tension on the water to retain it.

Soil texture is a primary determinant of how much water is held in a soil and under how much tension. Because water is held more tightly in small pores than in large pores and because coarse-grained soils have many fewer small pores than do fine-grained soils, soil texture has a pronounced effect on water retention (Figure 4.2). For example, coarse sandy soils have little ability to hold water under tension, so infiltrating water drains readily downward. Fine-grained

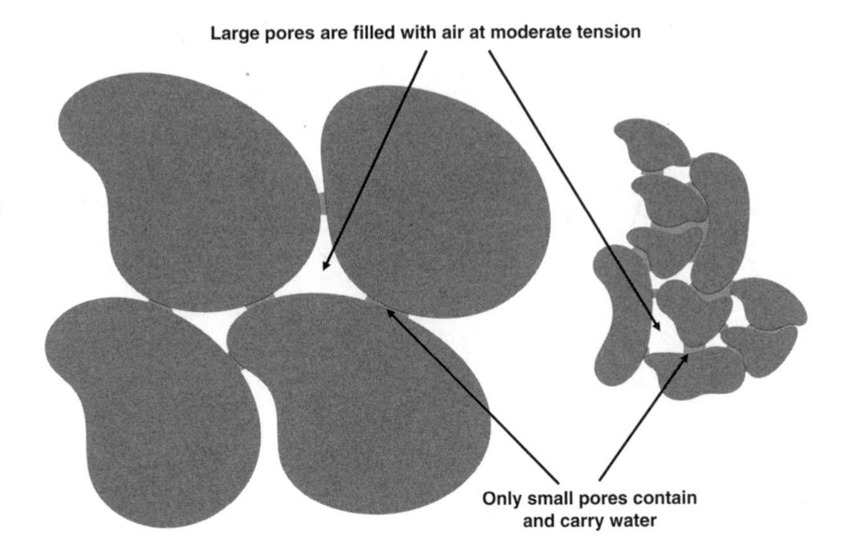

Figure 4.2 In an unsaturated soil, water is held in small pores while air fills the larger pores. The water is held under tension. The fine-grained soil (right) has more small pores than the coarse-grained soil (left), so the fine-grained soil retains more water.

soils, such as those with a significant clay content, will readily retain water under tension. Furthermore, as a soil dries out, the water moves into smaller and smaller pores and thus is held ever more tightly—or, as soil scientists say, is held under ever-greater tension.

Soils can vary in the amount of water they contain, from being totally saturated to being very dry with little residual moisture. The **volumetric moisture content** of a soil is a measure of the degree of saturation. The moisture content is typically represented by the Greek letter theta (θ) and is expressed as the volume of water per bulk volume of soil. Soil moisture varies from the soil surface to the **water table** (Figure 4.3). If all the openings in a soil are filled with water, the moisture content is maximum at the saturation value, θ_{sat}. For many soils, water will drain toward the **water table** until the moisture content reaches a level known as the **field capacity**, the volume of water per bulk soil volume, θ_{fc}, that is held against the force of gravity (see Figure 4.3).

Near the soil surface, plants exert a suction in their roots to extract the water held in the soil pores under tension. The water moves upward in the plant through conduits known as **xylems** and exits to the atmosphere (transpiration) through openings in the leaves known as **stomata** (Figure 4.4). As the plants remove moisture, the soil dries, and the water moves into smaller and smaller pores in which it is held under increasingly greater tensions. When the soil moisture content drops to θ_{wp}, the **permanent wilting point**, the plant can no longer exert a great enough suction to

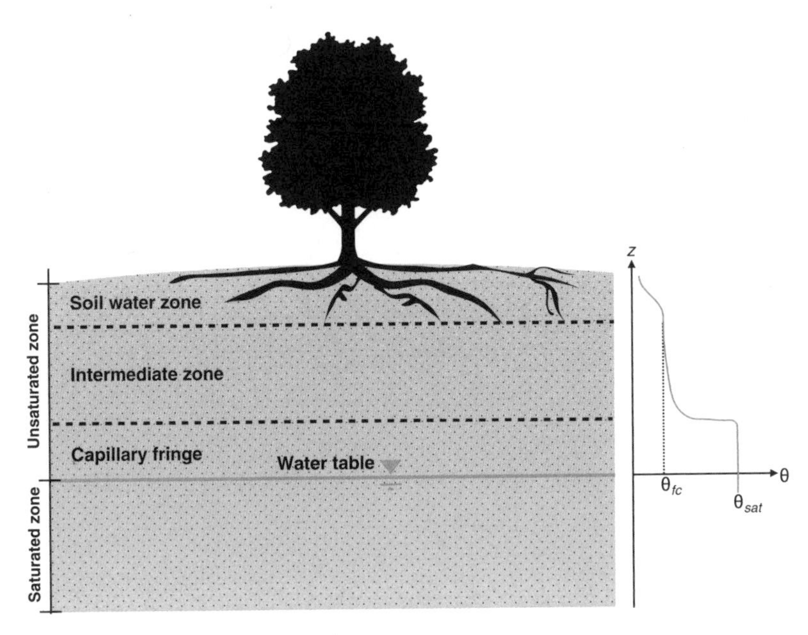

Figure 4.3 The distribution of moisture in the unsaturated zone. Water near the surface of the soil is available for uptake by plant roots. After several days of fair weather, the moisture content in this soil water or root zone decreases substantially due to evapotranspiration. Directly beneath the soil water zone (root zone), the moisture content tends to be constant over a substantial depth. The relatively constant value of moisture content in this region is referred to as the field capacity of the soil. The saturated zone above the water table is the capillary fringe. Redrawn from Hornberger et al. 2014.

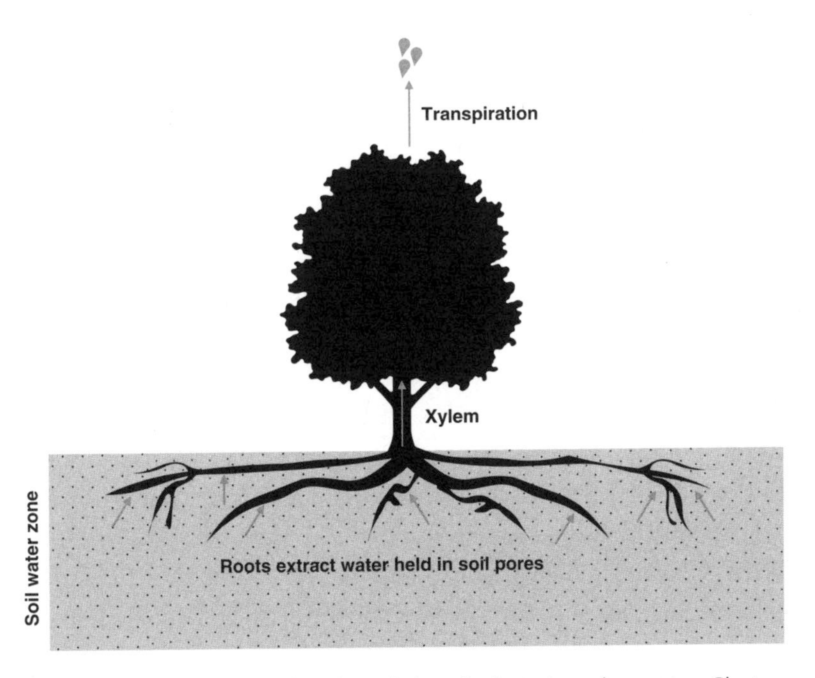

Figure 4.4 Schematic of water flow through the soil–plant–atmosphere system. Plants take up water through their roots. This water is transported through the xylem and released as water vapor to the atmosphere through the stomata (transpiration). Redrawn from Hornberger et al. 2014.

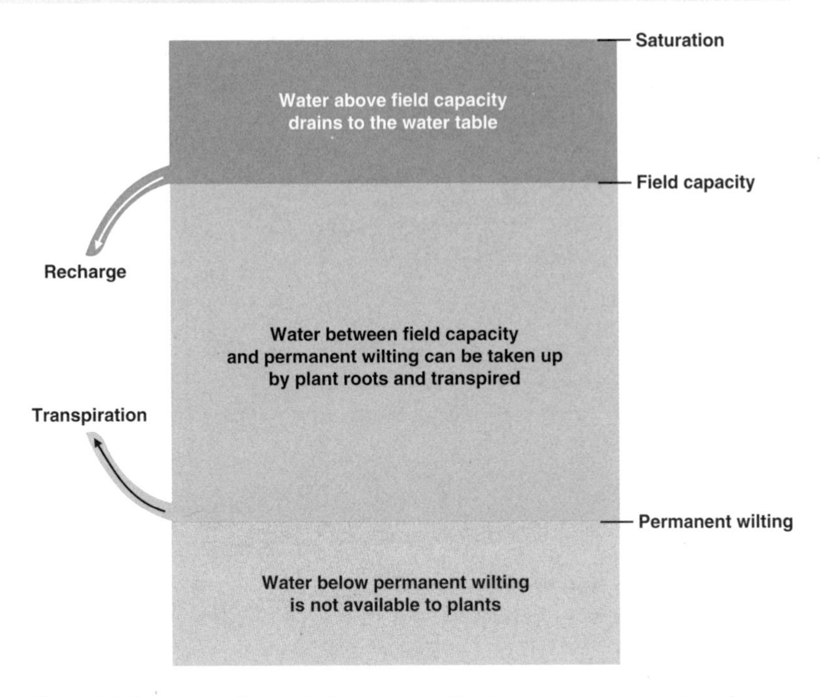

Figure 4.5 Schematic of water in the root zone. The top segment represents soil moisture from field capacity to saturation. Water above field capacity drains to recharge groundwater. The middle section represents water between field capacity and permanent wilting. This water is available to plants and can be taken up and returned to the atmosphere by transpiration. The bottom segment represents water below the permanent wilting point. Soil moisture below permanent wilting is not available to plants. Available moisture is the difference between field capacity and permanent wilting.

remove water from the soil, so the plant shuts down and stops transpiring—it wilts.

Water above the **field capacity** drains downward in soils and is unavailable to plants. Water held in soils below the **permanent wilting point** is also unavailable to plants. The **available moisture content**, θ_{aw}, is the difference between **field capacity** and permanent wilting (Figure 4.5). The best soils for plants have high available water, are not so coarse that they have low field capacities, and are not so fine that they have high **permanent wilting points**; for example, soils classified as loamy (Table 4.1) are typically good for plants. The **available water capacity (AWC)** of the root zone used in water balance calculations (Section 1.6) is based on θ_{aw}. For example, a sandy loam soil with $\theta_{aw} = 0.13$ and a root zone depth of 0.9 meters has an **AWC** of $(0.13\ m^3/m^3) \times (0.9\ m) = 0.117$ m or 117 mm.

Table 4.1 Typical soil water holding values as calculated using properties summarized by Clapp and Hornberger (1978).

Soil	Clay fraction	θ_{sat}	θ_{fc}	θ_{pw}	θ_{aw}
Sand	0.03	0.39	0.12	0.03	0.09
Sandy loam	0.09	0.43	0.25	0.12	0.13
Loam	0.19	0.45	0.33	0.17	0.16
Clay loam	0.34	0.47	0.38	0.23	0.15
Clay	0.63	0.48	0.41	0.29	0.12

Note: Field capacity is taken to be at a tension of 1/3rd bar and permanent wilting is taken to be at a tension of 15 bars. (One bar is standard atmospheric pressure at the surface of Earth.)

The water holding capacity of soils is affected by more than just the mineral grains of the soil. In particular, soil organic carbon affects θ_{aw} as well as the ability of soil to retain nutrients and interact with agricultural amendments of fertilizers, herbicides, and pesticides. The effect of additions of soil organic carbon on water retention is different for different soil textures, but in general leads to increases in θ_{aw}.

4.3 Soil Water Resources

The plant available water capacity of soils depends on the soil texture, soil organic matter, and on the depth of the root zone of soils. **Field capacity** and **permanent wilting point** can be estimated from soil texture and soil organic matter. Rooting depth can be estimated from vegetation coverage, so rooting depth implicitly incorporates climate and topography as well as soil texture information. **AWC** values, which are calculated from soil texture, soil carbon, and estimated plant root depths (Dunne & Willmott, 1996), vary from essentially zero, no water available for plants, to over 53 centimeters in areas with deep root zones and clay-loam soils (Figure 4.6). For a large fraction of land on which crops and forests grow, the **AWC** is in the range of 10 to 25 cm (100 to 250 millimeters). The **AWC** gives an indication of a maximum amount of water a soil can hold, but a key for plants is how much water is in the soil through time.

Plants control transpiration by regulating the opening of the **stomata**. When the **stomata** are open, water is lost through transpiration, and carbon dioxide, necessary for photosynthesis, enters. If a soil initially is at **field capacity**, transpiration will proceed at the maximum rate for prevailing conditions—that is, at the **potential**

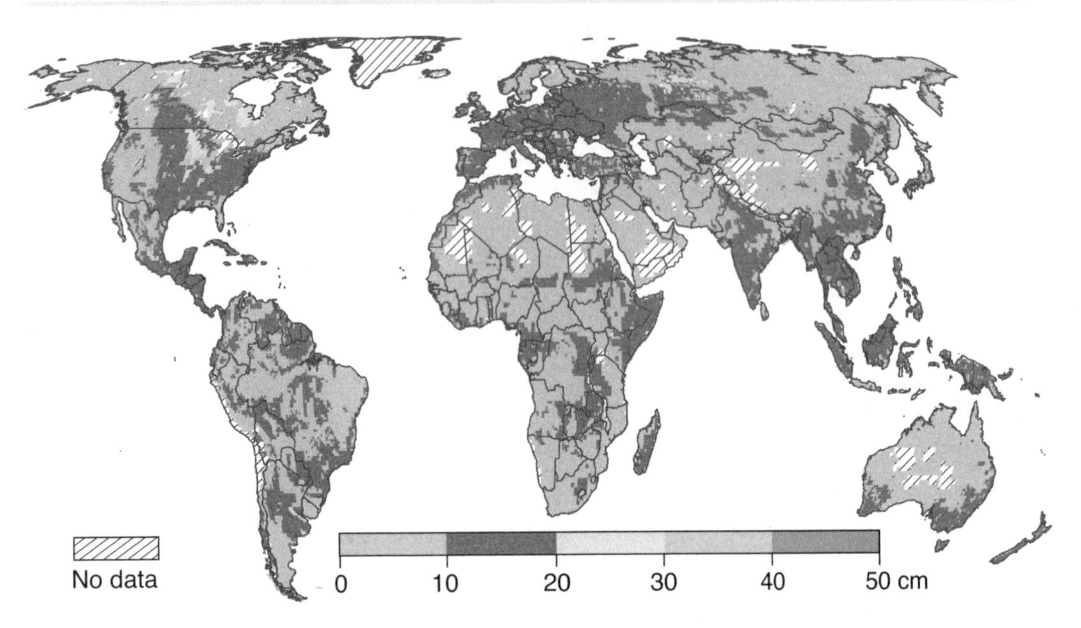

Figure 4.6 Plant available water over the rooting zone in centimeters of water. Data from Dunne and Willmott 1996.

evapotranspiration (PET) rate. The plants will draw water from the soil, thereby depleting soil moisture. Ultimately, if more water is not added to the soil, the soil moisture will drop to permanent wilting, at which point transpiration ceases. Even above permanent wilting, transpiration slows once the soil dries sufficiently.

When plants have to limit transpiration because of low soil moisture, they also lower photosynthesis. That is, transpiration and plant productivity are linked. The yield of crops can be strongly affected by soil water stress, decreasing by 50% or more relative to cases with optimal soil water conditions. Plants face a trade-off of keeping **stomata** open to maximize intake of carbon dioxide but using water rapidly versus partially closing **stomata** to conserve water but reducing the rate of photosynthesis. Some plants (e.g., maize) are referred to as **anisohydric. Anisohydric** plants pursue a course to keep **stomata** wide open until the soil moisture is close to permanent wilting; such plants use a strategy that prioritizes the uptake of carbon needed for plant metabolism but uses soil moisture at the **PET** rate at the risk of depleting available water. Other plants (e.g., wheat), referred to as **isohydric**, begin to close **stomata** once soil moisture reaches some critical level, θ^*, gradually dropping the rate at which transpiration occurs until perma-

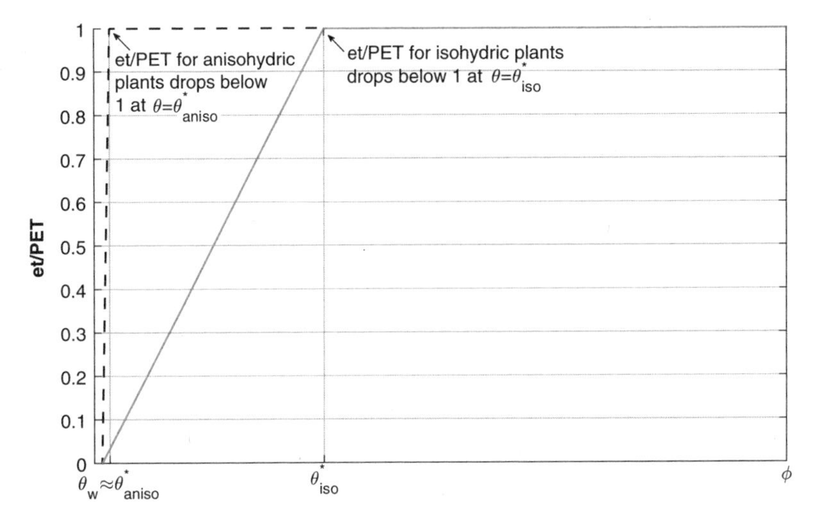

Figure 4.7 Relationship between the ratio of evapotranspiration (*et*) to potential evapotranspiration (PET) and soil moisture (θ) for isohydric (blue, solid) and anisohydric (black, dashed) plants. The value of θ is maximum (and equal to the porosity, Φ) at saturation. For soil moisture values greater than θ^*, the stomata are wide open and transpiration occurs at the potential rate (PET). As soil moisture decreases below θ^* the plant starts closing the stomata, thereby reducing *et*. At the permanent wilting point (θ_w), *et* is zero. Isohydric plants have a value of θ^* greater than anisohydric plants. Redrawn from Hornberger et al. 2014.

nent wilting is reached (Figure 4.7). These plants prioritize conserving water to avoid desiccation but can consequently face carbon starvation. One useful theoretical framework for estimating plant responses to increased severity of droughts due to climate change is the hypothesis that the drought mortality of **anisohydric** plants will be due to hydraulic failure and the drought mortality of **isohydric** plants will be due to carbon starvation.

The **water use efficiency (WUE)** of plants is the ratio of the biomass produced by plants to the amount of water transpired. For crops, the measurement is typically expressed as the mass of dry matter produced to the mass of water transpired. A related measure is the water productivity (WP), typically expressed as the mass of grain produced per area (kilograms/hectare) divided by the volume of water transpired per area (millimeters). **Anisohydric** plants should have an advantage over **isohydric** plants in terms of overall yield under conditions of moderate water availability, but they may not have an advantage in terms of water productivity. Future increases in food production will depend in part on plant breeding to increase plant **WUE**.

4.4 Soil Degradation and Water Resources

Determining the extent of soil degradation is very difficult. The time scale for the formation of soils is at least several thousand years. In contrast, the time scales associated with activities that degrade soils by increasing erosion rates, compacting the soil surface leading to lower water **infiltration** rates, depleting soil carbon or nutrients, and contamination by chemicals are much shorter, years to decades.

An example of an environmental catastrophe from the twentieth century is the "dust bowl" that occurred in the 1930s in the High Plains of the United States. Although it was recognized that the area was highly susceptible to erosion if the natural sod was disturbed, in the late 1800s and early 1900s, lands were divided into relatively small farm holdings. Ultimately plowing of the natural sod to grow wheat ensued. When a period of drought inevitably occurred, there was no natural sod to hold the soil in place. Great clouds of topsoil were carried away by winds, some deposited as far away as the East Coast of North America (Eagan, 2006). The area has since recovered from much of the ecological devastation, but a significant part of the recovery has relied on irrigated agriculture with the water being pumped from the High Plains (Figure 3.10) aquifer at an unsustainable rate (Chapter 3). If irrigation is curtailed, there is concern that severe wind erosion may occur again in the future.

In the 1980s, the Food and Agriculture Organization (FAO) of the United Nations conducted an extensive review and compiled estimates produced by many experts in soil science from around

Table 4.2 Percent of degraded land including categories from light degradation to very severe degradation. Data from Nachtergaele et al. 2011.

Region	Percentage of total area degraded
Sub-Saharan Africa	16.7
Middle East and North Africa	31.4
North Asia (east of Ural Mountains)	7.4
Asia and the Pacific	22.3
North America	5.2
South and Central America	18.5
Europe	25.9
World	16.8

the world. The GLASOD project (GLobal Assessment of human induced SOil Degradation) resulted in a detailed map of categories of soil degradation from light to very severe (Oldeman et al., 1990). An analysis of these results by country and region indicates that soil degradation is of serious concern (Table 4.2). For example, in India, with the population approaching 1.5 billion people, it is estimated that about 45% of the land is degraded, mainly by erosion (Bhattacharyya et al., 2015).

4.5 Concluding Remarks

The importance of a sustainable soil-water-plant system is hard to overstate. Archaeologists and historians have indicated soil erosion and loss of soil fertility to be primary reasons for the demise of the Roman Empire. For example, North Africa was an area that provided a great deal of grain to the Romans, but it ceased to be productive as soils eroded. Today many of these areas in Africa remain a desert. There are many other examples of soil erosion linked to the decline of civilizations (Montgomery, 2012). To sustain the needs of the growing world population it is essential that the joint resources of the soil-water-plant system be managed appropriately.

4.6 Key Points

- Openings in soils and rocks near the surface of the Earth typically are partially filled with water and partially filled with air. Hydrologists refer to this as the **unsaturated zone** or vadose zone. (Section 4.1)
- Water is held in the **unsaturated zone** by capillary forces that are inversely related to the size of the opening in which the water is held. Soils with fine texture have smaller pores and retain more water than coarse soils. (Section 4.2)
- The **volumetric moisture content** is a measure of water in soils. The moisture content is expressed as a fraction, the volume of water per bulk soil volume. The moisture content at saturation would be the porosity (Section 3.2) of the soil. (Section 4.2)
- Some water that infiltrates the soil percolates downward and recharges **groundwater**. The water that is retained in soil after **deep percolation** results in a moisture content known as **field capacity**. (Section 4.2)

- Plants exert a suction at their roots to extract water from the soil. Plant transpiration is the process whereby water is transported from the soil to the atmosphere through the **xylem**, channels within the plant. At the leaves the water exits to the atmosphere through **stomata**, openings in the leaves. (Section 4.2)
- When plants remove water from soils, the volumetric soil moisture drops, and the suction required to remove water increases. At a moisture content known as the **permanent wilting point**, the plant can no longer access water in the soil. (Section 4.2)
- The **available water capacity** of a soil is the difference between **field capacity** and permanent wilting. (Section 4.2)
- Plants face a trade-off of keeping **stomata** open to maximize intake of carbon dioxide and use water rapidly versus partially closing **stomata** to conserve water and reduce the rate of photosynthesis. (Section 4.3)
- The **water use efficiency (WUE)** of plants is the ratio of the biomass produced by plants to the amount of water transpired. Future increases in food production will depend in part on plant breeding to increase plant **WUE**. (Section 4.3)
- Soil degradation is a very serious problem worldwide. The strong interconnections within the soil-water-plant system condition both causes and impacts of soil degradation. (Section 4.4)

4.7 Example Problems

Problem 4.1. Evapotranspiration over a 3-month growing season from a field of corn with an area of 1 hectare (10^4 m^2) is 9,000 m^3. Take the effective soil depth to be 30 cm (0.3 m).

a. For an available water capacity of 0.15, what volume of water would be available for the 1 hectare field of corn if the soil were initially at **field capacity**?

b. If the 3-month growing season is taken to be approximately 90 days long, how many times at a minimum would the **AWC** have to be replenished to keep the crop from wilting?

Problem 4.2. Corn is an **anisohydric** crop. Wheat, on the other hand, is **isohydric**. Discuss how the water use and the number of times the **AWC** would have to be replenished might differ for a 1 hectare field of wheat relative to the field of corn in Problem 4.1.

Problem 4.3. Average soil production rates vary but have been estimated to be a few hundredths of a millimeter per year. Erosion rates on land dis-

Table 4.3 Time, soil depth, erosion rate, and production rate of an example soil.

T (years)	S (mm)	E (mm/year)	P (mm/year)
	300	0.64	0.06
	300	1.67	0.06
	500	0.64	0.06
	500	1.67	0.06

turbed by human activities have been estimated to be 10 or 100 times the production rate. The amount of time it would take to erode through a soil of depth S can be expressed as (Montgomery, 2007):

$$T = \frac{S}{E - P}$$

where T is time, S is the soil depth, E is the erosion rate, and P is production rate. Calculate the length of time to completely erode a soil under conditions given in Table 4.3; that is, fill in the first column of the table.

4.8 Suggested Readings

Eswaran, H., Lal, R., & Reich, P. F. (2001). Land degradation: an overview. In E. M. Bridges, I. D. Hannam, L. R. Oldeman, F. W. T. Pening de Vries, S. J. Scherr, & S. Sompatpanit (Eds.), *Responses to land degradation*. New Delhi: Oxford Press.

O'Geen, A. T. (2012). Soil water dynamics. *Nature Education Knowledge, 3*(6): 12.

4.9 References

Bhattacharyya, R., Ghosh, B. N., Mishra, P. K., Mandal, B., Rao, C. S., et al. (2015). Soil degradation in India: Challenges and potential solutions. *Sustainability (Switzerland), 7*(4), 3528–3570.

Clapp, R. B., & Hornberger, G. M. (1978). Empirical equations for some soil hydraulic properties. *Water Resources Research, 14*, 601–604.

Dunne, K. A., & Willmott, C. J. (1996). Global distribution of plant-extractable water capacity of soil. *International Journal of Climatology, 16*(8), 841–859.

Eagan, T. (2006). *The worst hard time: The untold story of those who survived the great American dust bowl*. Boston: Houghton Mifflin.

Hornberger, G. M., Wiberg, P. L., Raffensperger, J. P., & D'Odorico, P. (2014). *Elements of physical hydrology* (2nd ed.). Baltimore: Johns Hopkins University Press.

Montgomery, D. R. (2007). Soil erosion and agricultural sustainability. *Proceedings of the National Academy of Sciences of the United States of America, 104*(33), 13268–13272.

Montgomery, D. R. (2012). *Dirt: The erosion of civilizations.* Berkeley: University of California Press.

Nachtergaele, F. O., Petri, M., Biancalani, R., van Lynden, G., van Velthuizen, H., & Bloise, M. (2011). *Global Land Degradation Information System (GLADIS). An information database for land degradation assessment at global level.* http://www.fao.org/nr/lada/index.php?option=com_docman&task=doc_download&gid=773&lang=en

Oldeman, L. R., Hakkeling, R. T. A., & Sombroek, W. G. (1990). *Global assessment of soil degradation.* Wageningen, the Netherlands: ISRIC.

II

Demand-Side Sectors

CHAPTER 5

Agricultural Water Use

5.1 Introduction

There are many demands on our water resources. A series of simple questions captures some of the basic information we need in understanding how water is used and, consequently, how water may be allocated under scarcity. (1) What users place the largest demands on our water resources? (2) What trade-offs are prominent when those users are making decisions? (3) How do water resource trade-offs fit into the portfolio of broader trade-offs for each of the users? These questions—linked to the diverse users of water often referred to as demand-side sectors—play a role in how we manage our water resources. In the next few chapters, we will look at four key demand-side sectors: agriculture, energy, households (i.e., domestic), and the environment. We will explore the many trade-offs those sectors face—reliability, economics, the environment, and water use—to provide us with the foundation necessary to explore future opportunities in water management.

In considering water uses, we can distinguish between water withdrawals and water consumption. Water withdrawals include surface water or **groundwater** removed from its source. Water consumption is the portion of the water withdrawn and lost through evaporation or transpiration. The portion of the water withdrawn, but not consumed, is eventually returned to a surface water or **groundwater** source, and it is available for other downstream uses. The portion of the water consumed (i.e., evapotranspired) is no longer available for other uses in the proximate area.

Globally, the largest demand-side sector of water withdrawals is agriculture. Agriculture relies heavily on healthy soil, which is also important for storing water for crops (Chapter 4). Soil is a living ecosystem with a range of micro- and macro-organisms that help convert organic matter and minerals into nutrients used by crops. Healthy soil ecosystems retain nitrogen and other nutrients for crops, uphold soil structure, protect against pests and diseases, and act to retain soil moisture. Healthy soil ecosystems are a fundamental component to our agricultural-water systems because they maintain soil moisture on cultivated lands and improve drought resistance.

Agricultural decisions, such as what type of agricultural procedures to adopt or what type of irrigation to adopt, must consider reliability, economic costs, and environmental impacts, in addition to water requirements. In short, there are often a number of tradeoffs that can indirectly affect water requirements, as well as tradeoffs that can directly affect water requirements.

5.1.1 Types of Agriculture

A spectrum of agricultural types can be used to manage soil and soil water. **Conventional agriculture** depends highly on off-farm inputs—and thus large capital investments. **Conventional agriculture** boasts high efficiency in labor, but it requires investments in mechanical devices such as tractors to plow and cultivate the land and irrigation systems to transfer water to fields for crops. **Conventional agriculture** tends to focus more on high-yielding monoculture crops and less on sustaining an equilibrium in soil ecosystems (Figure 5.1) (Rosset & Altieri, 1997). To maintain large yields without pests and weeds, this type of agriculture uses fertilizers, pesticides, and herbicides. **Conventional agriculture** in the United States is associated with the goal of increasing yields, a movement often linked to an increasing dependence on inputs such as fertilizers, pesticides, and herbicides. **Conventional agriculture** tends to be at large scale because the associated management techniques make it possible to farm large tracts of land. Nevertheless, some small farms also apply conventional techniques.

In addition to **conventional agriculture**, we can consider **conservation agriculture** and **organic agriculture**. **Conservation agriculture** is characterized by little to no **tillage**, the use of cover crops, and crop diversification to sustain soil ecosystems (Figure 5.2).

Tillage is the process of breaking up and mixing topsoil to destroy weeds and pests and to redistribute and release nutrients from past harvest residues (Figure 5.3). **Tillage** can be done with human and animal labor, but because it is very labor intensive, farmers with

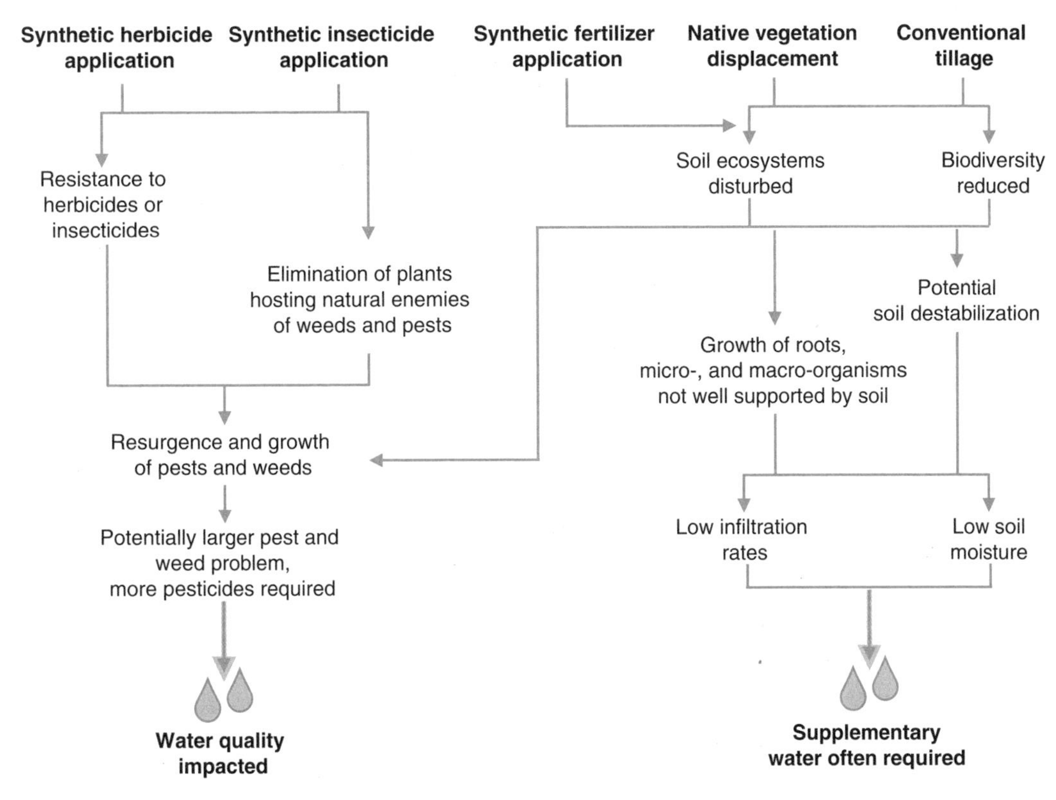

Figure 5.1 Potential management challenges facing conventional agriculture due to mechanization and the application of fertilizers, herbicides, and pesticides. Over the long term, the application of synthetic pesticides can allow pests and weeds to build resistance to pesticides. Synthetic fertilizers, monoculture, and tillage can degrade soil quality. The application of synthetic additives can impact water quality if not maintained properly. Soil compaction and destabilization can result in lower rates of water infiltration. Without cover crops, soil erosion can be increased.

access to machinery often employ tractors. Tractors with plows can be used to break up the soil, incorporate residues, and sow seeds. The major advantages of a **tillage** system over a no-till system is that it reduces reliance on herbicides, promotes consistent seed planting depths, and kills weeds or pathogens in the soils (Peigné et al., 2007). The major disadvantages of **tillage** systems are the increased susceptibility of the soil to erosion, the loss of soil moisture retention capacity, and compaction of soils from the plow or machinery. **Tillage** also aerates the soil, accelerating the carbon cycle, which increases decomposition and releases carbon dioxide.

Over the long term, **tillage** can degrade soil quality, requiring increasing amounts of fertilizer input to maintain yields. Degradation of soil quality from **tillage** is acute in tropical climates, where the warm temperatures accelerate decomposition processes. In some

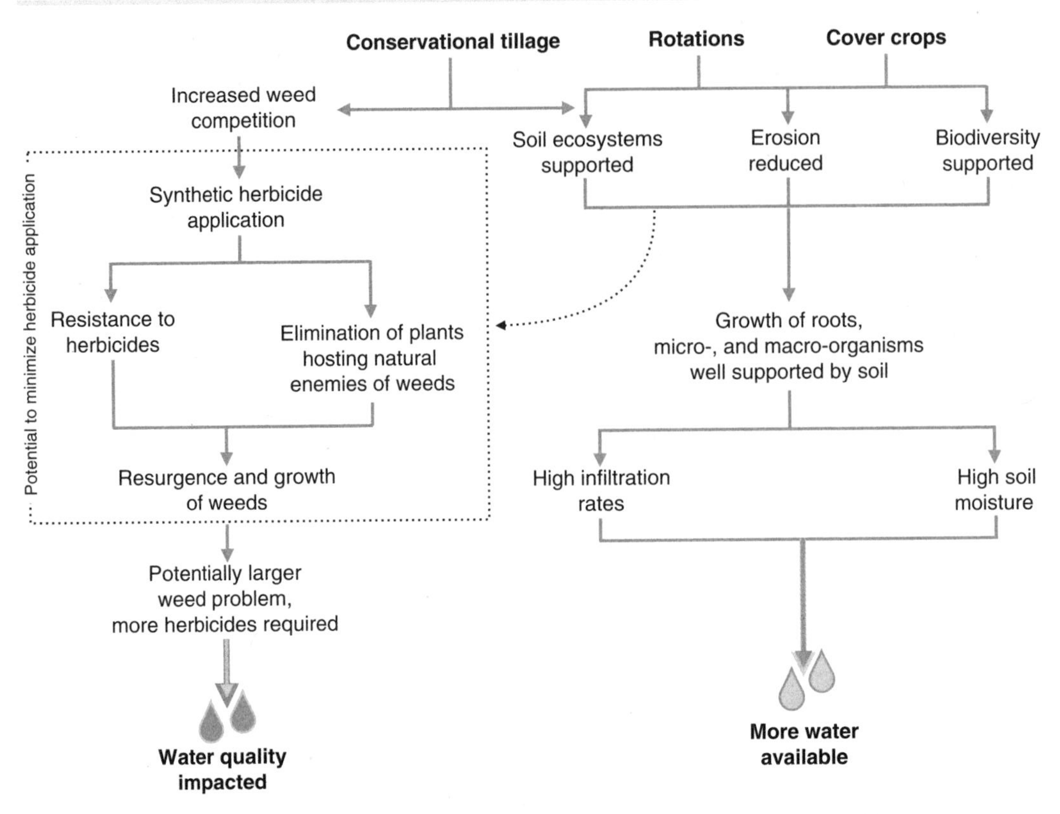

Figure 5.2 Potential management benefits and challenges facing conservation agriculture. Diversification of crops through rotations and the use of cover crops can build a strong soil ecosystem. Strong soil ecosystems promote water infiltration and soil water storage. In wet climates or areas with poorly drained soils, the increased soil infiltration can lead to waterlogging; in dry climates, it is likely that more soil water is available for crops, reducing the requirement for irrigation. In some cases, the reduction of tillage can lead to an increase in weeds, requiring herbicides. The application of herbicides, if managed poorly, can lead to water quality issues or more severe weed problems. Crop diversification and cover crops provide many benefits, such as healthy soil ecosystems, which may be able to minimize the need for herbicide application.

areas of the world, weeds and pests are destroyed with planned burnings followed by **tillage**. Burning residues can produce short-term benefits but also can degrade the soil over time because the soil ecosystem is damaged.

Tillage practices fall on a spectrum from conventional to no till. *Conventional tillage* is defined as cultivating a crop with less than 30% residue remaining. The residue is the material from crops that are not harvested; the residue acts as a mulch, protecting the soil from wind, rain, and sun, and creating habitat for beneficial insects. Conservation tillage leaves 30% or more residue on the surface (Soil Science Glossary Terms Committee, 2008). *No-till systems* only mix the surface layer (see Figure 5.3), leaving residue from

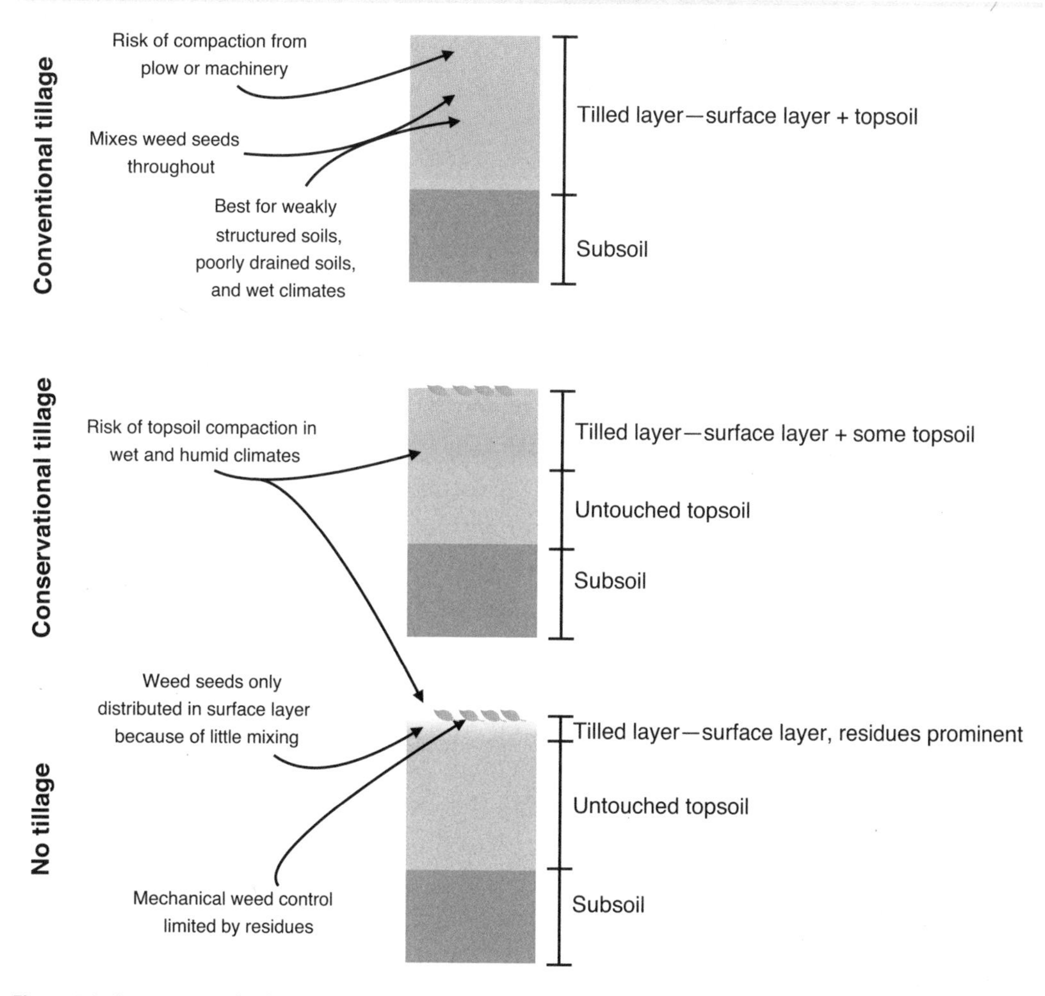

Figure 5.3 Comparison of soil layers in conventional tillage, conservational tillage, and no-tillage systems. Each system has potential management benefits and challenges.

previous harvests on the soil such that seeds are planted directly into a small slot in the land surrounded by residue; no-till systems are also referred to as direct seeding systems (Soil Science Glossary Terms Committee, 2008). In no-till systems, the soil is not disturbed or is disturbed very minimally. To remove weeds, herbicides are applied optimally to residues, with a focus on maintaining the equilibrium in soil ecosystems. The major advantages of no-till systems are reductions in soil erosion, maintenance of healthy soil ecosystems, conservation of soil moisture, and a reduction in external inputs such as fuel for tractors or physical labor.

Cover crops, such as soybeans and winter rye, are grown to protect soil fertility, reduce erosion, control pests and weeds, and recycle nutrients during fallow seasons or fallow years of crop rotations. Although many cover crops are not typically intended for harvest, some cover crops can be used as forage for livestock. Cover crops typically add nitrogen to the soil in a form that other crops can use, which can reduce the need for synthetic fertilizers. Cover crops add diversity to cropland, increasing the diversity of microbes in the soil and encouraging beneficial insects to thrive, ultimately protecting against disease and pests. Roots from decomposing cover crops can increase water **infiltration**. Residues from cover crops act like a protective layer on top of the soil, reducing the impact of rain. Without this protective layer, the soil surface can erode, crust, or seal, ultimately reducing the **infiltration** capacity of the soil.

To integrate cover crops into an agricultural system, a cropping system that focuses on crop diversification is used. Crop diversification can be embraced, for example, by using crop rotations or intercropping. Whereas monocropping focuses on the same crop throughout the field every season, *crop rotation* is the practice of alternating diverse crops throughout many years on one plot of land. For example, corn might be grown in the first year, oats grown in the second year, and then the field might be used for pasture for several years before starting the rotation again. Not only does crop rotation promote pest and weed resistance, also it promotes plant rooting at a range of depths. Crop diversification leads to a high diversity of roots and rooting depths, which provides for better water and nutrient distribution in the soil. There is evidence that increasing crop diversity through the use of crop rotation improves yield stability, especially during drought conditions (Gaudin et al., 2015).

In some areas, *intercropping* can be used so that two or more crops are grown at the same time on the same plot of land. One major disadvantage of intercropping is the competition among crops for sunlight, water, and nutrients. As a result, intercropping requires knowledge about each crop and how the crop interacts with other crops to ensure effective integration between or among them.

Many of the central tenants—little **tillage**, cover crops, and crop diversification—of **conservation agriculture** and diversified farming systems can be practiced with **organic agriculture**. Organic agriculture is perhaps the most well-known certification for alternative agricultural practices. There are many certification agencies for

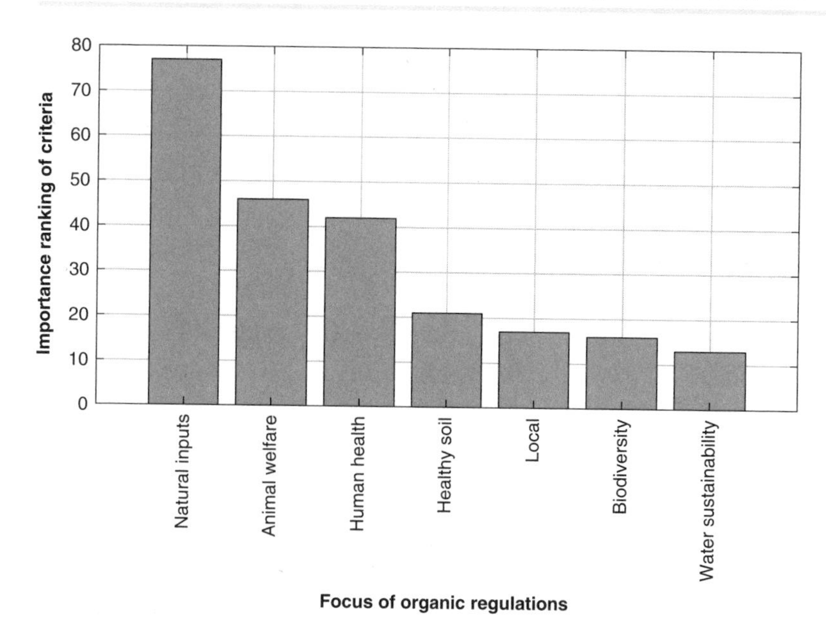

Figure 5.4 Importance ranking for seven criteria across national and international regulations for organic agriculture. Organic agriculture has well-defined certifications, which can differ slightly by certifying agency or country. Most regulations for organic certifications focus on natural inputs, animal welfare, and human health. Water sustainability is the least important criterion across a selection of national and international regulations for organic agriculture. Data from Seufert et al. 2017.

organic agriculture, at both the national and international scales. Organic regulations for certifications either (1) define natural inputs and natural practices that are allowed or recommended for use in **organic agriculture** or (2) define synthetic inputs and industrial practices that are prohibited for use in **organic agriculture.**

One of the major criticisms of modern-day **organic agriculture** is its weak emphasis on promoting environmental sustainability through holistic approaches for soil and water management compared to its strong emphasis on avoiding synthetic inputs. Recent research reviewing organic regulations from a selection of national and international agencies found that both soil and water management ranked much lower in regulatory importance than natural inputs, animal welfare, and human health (Figure 5.4). As a result, there has been a movement toward using local certifications that more adequately capture environmental sustainability through holistic approaches.

Two additional types of agriculture that are becoming increasingly popular include diversified agriculture and sustainable agriculture. The emphasis of *diversified agriculture* is on multiscale biodiversity. Diversified farming promotes diversity of plants and animal species on the farm as well as in the region. *Sustainable agriculture* puts an emphasis on social, environmental, and economic equity. Sustainable farming prioritizes many of the practices fundamental to conservational, organic, and diversified farming through the lens of equity.

In short, farming practices that focus on healthy soil ecosystems also increase the capacity of soil to retain and release water optimally for crop growth. Residues left on croplands from conservation **tillage** or cover crops protect the soil from direct raindrop impact, which can seal the top layer of the soil. Clogged pore space, combined with heat from high temperatures or the sun, can result in a surface crusting that reduces **infiltration** and increases runoff. Residues left on croplands reduce soil temperatures and water evaporation from the soil. The reduction in evaporation can promote water conservation, which can reduce the requirement for irrigation water and energy to distribute that water, thus saving money. Activities from macro-organisms and decomposing roots from previous harvests create pore space in the soil, allowing water to infiltrate more easily. High levels of organic matter can increase the water holding capacity of the soil, ultimately increasing the storage of water in the soil. Although conservation **tillage** and residues are beneficial in dry and hot climates, there is some evidence suggesting that conservation **tillage** and residues can lead to **waterlogging**, or overwatering, in cool and wet climates. Consequently, conservation tillage in cool and wet climates can lead to declines in yields.

5.2 Rainfed and Irrigated Systems

Crops in some locations can thrive using only water from natural precipitation. *Rainfed agricultural systems*, as the name implies, rely on natural precipitation to supply crop water requirements. Dryland farming is rainfed agriculture in regions where precipitation varies throughout the year; this results in crops facing water stress during part of the growing season.

Irrigated agricultural systems use supplementary water supplies from collected and stored rainwater reserves, surface water, **groundwater**, or recycled water to provide needed water to crops. Deficit irrigation and limited irrigation are two approaches to

Table 5.1 Comparison of energy requirements, cost, water supply application, efficiency, and potential water losses for surface irrigation, sprinkler irrigation, and microirrigation systems. Data from Amosson et al. 2011.

	Surface irrigation	High-pressure sprinkler irrigation	Microirrigation
Examples of types	Flood, furrow	Center pivot irrigation, spray irrigation, hanging pipe spray irrigation	Drip irrigation, subsurface drip irrigation, low-pressure sprinkler
Energy requirement at field scale	Usually not mechanized; uses gravity to spread water across fields using soil as a conduit	Mechanized; uses motors to pump water through attached, high-pressurized sprinklers	Mechanized; uses motors to pump water through pipes with holes that are strategically placed along or buried under crops to focus water delivery on the root lines
Cost	Low	Moderate	High
Water supply application	Large applications, intermittently	Small applications, continuously	Small applications, continuously
Efficiency	Limited amount of precision; generally considered least efficient	Can be linked to precision applicators to increase efficiency; generally considered midrange	Often linked to precision applicators to increase efficiency; generally considered most efficient
Potential water losses	Evaporation, deep percolation, runoff	Evaporation, wind drift, deep percolation	Little to none if managed correctly

increasing water-use efficiency in irrigation. *Deficit irrigation* eliminates irrigation during the crop-growth stages that are less impacted by water stress. Here we focus on four stages of crop growth: initial stage, where the crop establishes itself; crop development stage, where the crop is vegetative; midseason stage, where the crop flowers and yields form; and late season stage, where the crop ripens and is harvested. The midseason stage is most vulnerable to water stress, and the crop development and late season stages tolerate modest water stress (Brouwer & Heibloem, 1986). Deficit irrigation would, for example, target water conservation during the development stage. *Limited irrigation* systematically reduces water throughout the entire growing season; limited irrigation is best used when combined with rainfed systems in wet or humid climates.

The three main types of irrigation systems include surface, sprinkler, and microirrigation (Table 5.1). **Surface irrigation** systems spread water across the surface of crop fields, applying water at the highest parts of a field so that it flows down the slope and across the field. Water can be spread through flooding the entire field so that it ponds at the surface or by directing the water into furrows or small channels.

Sprinkler irrigation systems spray water through sprinklers using a high-pressure system that provides water from the top downward. Center pivot irrigation rotates the sprinklers or nozzles around a pivot point, irrigating a circular pattern of crops. If you have ever flown over the Midwest in the United States, it is likely that when you looked out your window, you saw large green circles atop the brown landscape; these are plots of land that use pivot irrigation systems. Linear-move irrigation, where sprinklers move forward and backward rather than circularly, can be used for rectangular fields. Spray irrigation is best in flat terrains where surface runoff can be minimized. Low-drift nozzles, or nozzles that emit larger droplets of water than traditional nozzles, can minimize water losses from wind drift or evaporation (Figure 5.5).

Microirrigation systems slowly apply small volumes of water to the area surrounding a crop's root zone; this can be done with both drip and sprinkler technologies. There are two forms of drip irrigation: surface drip or subsurface drip. Surface drip systems distribute water using low pressure distribution at the surface above the root zone. Subsurface drip systems apply water underground to the root zone directly.

Figure 5.5 Water gains and losses in a crop system.

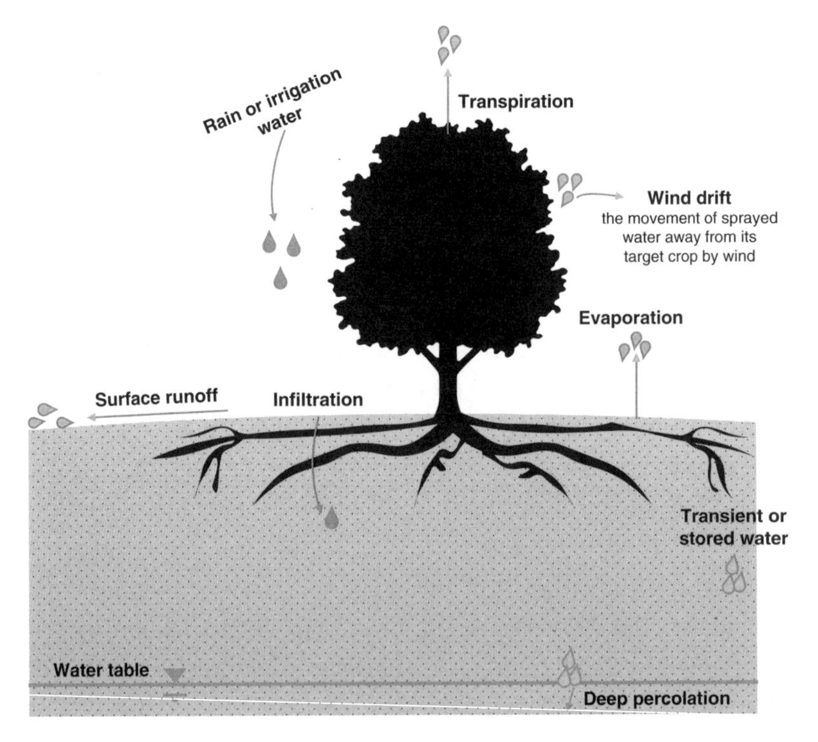

Table 5.2 Conveyance efficiency ranges for various types of conveyance systems. Data from Brouwer et al. 1989.

Types of conveyance systems	Conveyance efficiency (E_c)
Earthen canal[a]	
Short (<200 m)	From ~80% (sandy soils) to ~90% (heavy clay soils)
Medium (200–2,000 m)	From ~70% (sandy soils) to ~85% (heavy clay soils)
Long (>2,000 m)	From ~60% (sandy soils) to ~80% (heavy clay soils)
Lined canal[b]	~95% (lined with bricks, plastic, or concrete)
Pipeline[c]	~100% (assuming no leaks in the system)

[a] The short, medium, and long earthen canals have a range of values based on the material of the canal. Because water infiltrates more easily through sand than clay, more water is lost during conveyance when a canal is constructed from sand than from clay.
[b] Lined canals can reduce infiltration, but because the canals are not covered, water loss still occurs from evaporation.
[c] Pipelines have the highest efficiency, but are subject to cracks and leaks.

Each type of irrigation system has pros and cons (see Table 5.1). The selection of irrigation system depends on climate, crops, and available labor and capital. For an irrigation system to work best for its application, it must be designed, implemented, and managed properly.

When rainwater or irrigation water is applied to a field, water can be lost through evaporation, wind drift, or runoff (see Figure 5.5 and Table 5.1). Only a portion of the water that infiltrates is stored in the root zone for use by crops. Some water that infiltrates is lost to **deep percolation**. Water also can be lost before it is applied to a field during conveyance from the water source to the agricultural field.

Irrigation efficiency (E, %) is a composite of water conveyance efficiency (E_c, %) and **water application efficiency** (E_a, %). **Water conveyance efficiency** takes into account the transmission losses of water as it is conveyed from the water source to the agricultural field through an earthen canal, lined canal, or pipeline (Table 5.2). Losses can be a result of water seepage, spills, leaks, and evaporative losses. **Water conveyance efficiency** is the ratio of the volume of water that gets to the field (V_f) to the total volume of water diverted to the agricultural field from the water source (V_d).

$$E_c = \frac{V_f}{V_d} \times 100$$

The objective of irrigation is to supply water to the crop via soil water in the root zone. The **water application efficiency** (E_a, %)

Table 5.3 Application efficiency ranges for the three main types of irrigation. Data from Irmak et al. 2011.

Types of irrigation	Application efficiency (E_a)
Surface	~40%–80%
Sprinkler	~65%–90%
Microirrigation	~80%–95%

Note: Irrigation efficiency can vary due to climatic factors, type of subsystem, management of irrigation systems, and local factors like soil.

provides an indication of how effective the application of irrigation is at getting to and staying in the root zone (Table 5.3); thus, the **water application efficiency** is the ratio of the volume of irrigation water stored in the root zone (V_s) to the total irrigation water applied to the field (V_f).

$$E_a = \frac{V_s}{V_f} \times 100$$

Irrigation efficiency (E) is the overall efficiency of the physical irrigation system; this includes water conveyance and application.

$$E = \frac{E_a}{100} \times \frac{E_c}{100} \times 100$$

or we can simplify the equation to

$$E = \frac{V_s}{V_d} \times 100$$

To maximize the benefits from irrigating, it is important to calculate both the quantity of water required for crops and the timing of irrigation. This can be done using irrigation scheduling. Irrigation scheduling helps farmers avoid crop water stress as well as soil **waterlogging**. Irrigation scheduling requires fundamental knowledge about weather conditions, stages of crop growth, type of crop roots, soil water capacity, and irrigation techniques. An understanding of weather conditions can promote the efficient use of rainwater and reduce irrigation requirements. As we learned with deficit irrigation, limiting water during certain stages of crop development can be a useful strategy when trying to conserve water resources.

Not all crops require the same amount of water to beat water stress, and crop roots can play a vital role in determining the active root zone and water storage available. Nor do all soils hold water equally—finer-grained soils hold more water than coarser-grained soils (Chapter 4).

5.3 Reliability

5.3.1 Type of Agriculture

One way to measure the reliability of different agricultural systems is by estimating yields in terms of crop matter harvested per unit of cultivated land. When comparing yields from conventional and **organic agriculture**, it is critical to keep in mind that yields can be compared from the crop perspective (i.e., crop yields per season, or crop matter harvested per cultivated land per season) or from the agricultural systems perspective (i.e., crop yields over the full crop rotation period or crop matter harvested per land cultivated per rotation).

On the scale of crops, the yields using organic practices are, on average, about 67% to 81% of the crop yields using conventional practices (Fedoroff et al., 2010; Kniss et al., 2016; Ponisio et al., 2015; de Ponti et al., 2012; Seufert et al., 2012). Because **organic agriculture** emphasizes natural fertilizers, the crop yields may be limited by nutrients. On the systems scale, conventional agricultural yields also tend to look different from organic agricultural yields (Connor, 2013) because **organic agriculture** relies heavily on crop rotations and these crop rotations can be very diverse.

The benefits of crop diversity and conservation **tillage** are especially pronounced during hot and dry conditions. The yield benefits of crop diversification and conservation **tillage** appear less pronounced during cool and wet conditions, with some evidence that conservation **tillage** in wet areas can lead to **waterlogging** of the soil and reduced yields. The complexity in crop yields discussed here highlights the critical need to adopt the best management practices for the local context, informed by the regional and global context.

5.3.2 Rainfed and Irrigated Systems

Each crop has a *yield potential*, which is the maximum yield that can be attained under perfect conditions—solar radiation, temperature, and water—given the crop's features. For rainfed crops, adequate water may not be reliable or consistent, so the crop yields may be water limited. The maximum yields under water-limited rainfed systems are less than the yield potential. Crops can also be nutrient limited and be impacted by weeds, pests, diseases, and pollution, reducing yields further. The difference between yield potential and the actual yield is the *yield gap*.

Irrigation can increase yields and reduce **interannual variability** of production because water is often reliable year-round and can be delivered during the crop's most vulnerable growth periods. Generally speaking, crop productivity for irrigated land is twice that of rainfed land, and production variability for irrigated land is half that of rainfed land (Tubiello et al., 2002).

5.4 Economic Trade-offs

5.4.1 Type of Agriculture

Even though organic yields are lower than conventional yields at the crop and system levels, products of **organic agriculture** are sold at a premium price. Current premiums are about 30% on organic agricultural crops (Crowder & Reganold, 2015). As a result of the premium price, organic agricultural products are more profitable than products from **conventional agriculture**. Nevertheless, the conversion to certified **organic agriculture** holds many challenges for smallholder farmers. Economists sometimes use the ratio of monetized benefits (e.g., soil health, biodiversity) to costs (i.e., labor costs, equipment costs, energy costs) to compare two or more options. An analysis comparing many different studies showed that the benefit to cost ratio was higher for **organic agriculture** than **conventional agriculture** at both the crop and system levels (Crowder & Reganold, 2015).

Organic agriculture is certified, likely promoting the premiums that allow **organic agriculture** to be more profitable than **conventional agriculture**. Because **conservation agriculture** does not have a certification, it is less likely that crops harvested under these practices would gain the premiums seen for organic farming. There is evidence that some elements of conservational farming can maintain similar profits as conventional farming. For example, increasing cropping diversity through longer rotations has been shown to promote yields while decreasing the use of agrichemicals.

The result of increased yields and decreased inputs has allowed crop diversity via rotations to maintain similar net profits to conventional rotations while providing environmental benefits such as reduced energy use and reduced water pollution.

5.4.2 Type of Irrigation

Picking the most appropriate irrigation system can be complicated and involves a number of trade-offs. Agricultural factors such as the crop mix, labor availability, field size, and climate are important to consider. Irrigation factors like the availability of water, the water source, the application efficiency, the conveyance efficiency, and economic factors play a role in choosing appropriate technologies. The financial investments for an irrigation system vary significantly for the three major types of irrigation systems: surface, sprinkler, and **microirrigation.** Of the three systems, the gross investment—or the capital costs of the irrigation system not including the well, pump, and engine—is often highest for **microirrigation** systems. The gross investments are often lowest for **surface irrigation** systems.

Each irrigation system also requires annual operating costs. More efficient irrigation systems (e.g., **microirrigation**) can reduce pumping costs, field operation costs, and field labor. Consequently, for some water- and crop-use scenarios, the total benefits from investing in more costly irrigation systems may be higher than the total costs.

5.5 Environmental Trade-offs

Agriculture contributes to emissions of greenhouse gases (Chapter 10), which lead to climate change (Figure 5.6). Directly, agriculture is responsible for about 15% of global greenhouse gases, of which a majority are related to animal agriculture (Grosso & Cavigelli, 2012). If land-use changes resulting from agriculture are incorporated, **greenhouse gas** emissions from agriculture increase another 7%. Consequently, the approach that society takes to farming and the crops that society prioritizes for harvest play important roles in managing greenhouse gases.

Agricultural intensification through the applications of synthetic fertilizers, pesticides, and herbicides, fossil-fueled equipment, and genetically modified seeds have allowed some high-income countries to reforest previously cropped lands, acting as **greenhouse**

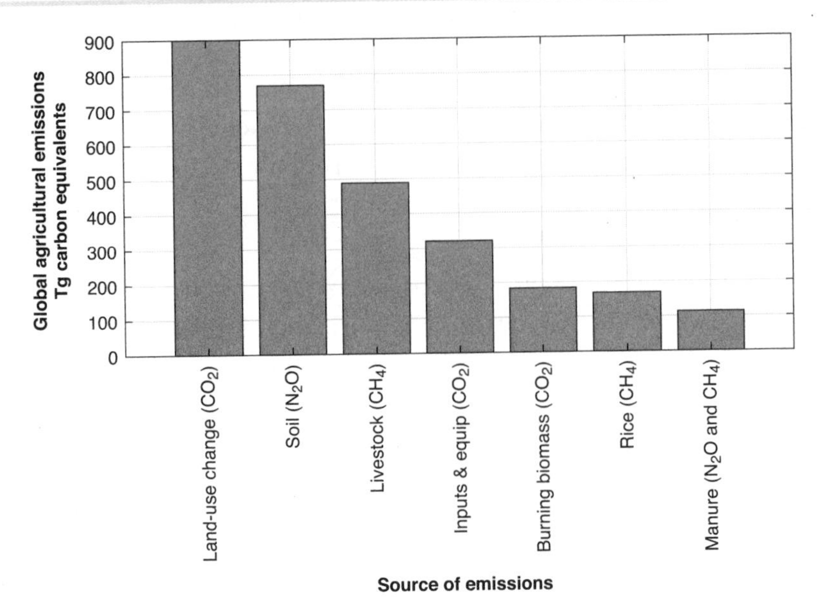

Figure 5.6 Global sources of agricultural emissions in 2010 shown as Tg carbon equivalents, where a Tg is 1×10^{12} g. Globally, agriculture is responsible for about 22% of greenhouse gas emissions. Data from Grosso and Cavigelli 2012.

gas sinks. At the same time, moving toward many of the alternative forms of agriculture discussed here can, for example, reduce **greenhouse gas** emissions from mechanized equipment and synthetic inputs. Many alternative forms of agriculture also promote climate resilient farming practices: little to no **tillage** promotes soil health, cover crops promote soil health and maintain soil moisture, and crop diversification reduces the impact of pests and weed growth.

The energy requirements for organic and conventional crops vary, depending on the type of energy: direct or indirect. *Direct energy* is energy used on the farm, and it includes labor, fuel, electricity, and equipment. *Indirect energy* is energy consumed for the manufacturing and distribution of agricultural inputs such as fertilizers, pesticides, herbicides, and seeds. Although direct energy consumption tends to be higher for crops grown under organic practices, some studies have reported higher direct energy for crops grown under conventional practices due to factors such as crop type, season length, and on-farm production practices. Indirect energy requirements for **conventional agriculture** are higher than for **organic agriculture**; **conventional agriculture** relies heavily on fertilizers and pesticides, and the manufacturing of these products is very energy intensive. Similar findings apply to practices of crop diversification: direct energy is reduced substantially on farms with more diverse rotations, even though labor use is greater.

Many of the best management practices for all agricultural types reduce the requirement for fertilizers and pesticides, which can reduce the impact of agriculture on water quality. There is evidence that increasing cropping diversity through the use of longer crop rotations reduces fertilizer and pesticide applications while maintaining yields and keeping weeds at bay—as a result, freshwater toxicity is reduced greatly. The evidence suggests that more diverse cropping systems, with longer rotation periods, can reduce agrichemical applications, with fertilizers and pesticides tuned to the cropping system rather than driving the cropping system.

5.6 Water Trade-offs

The amount of water required for crop growth depends highly on the climate and the crop. Temperature, humidity, wind speed, and solar radiation are critical climate factors. If the same crop was planted in both a hot, dry, windy, and sunny environment and a cool, humid, and cloudy environment, the water requirements would be higher for the former. The crop type and the stage of growth both affect the water required and the plant's vulnerability to water stress. For example, plants in the midseason stage are most vulnerable to water stress, whereas plants in the crop development stage are better able to tolerate water stress. In the initial stage, it is important that the soil near the root zone be wet to ensure proper root development.

The **crop water need** (ET_{crop}) is the depth (in millimeters) of water needed to replace the water consumed, that is, the water evapotranspired. To estimate this value, we use the reference crop evaporation (ET_o) and the crop factor associated with the crop of interest (K_c):

$$ET_{crop} = ET_o \times K_c$$

The ET_o is the evapotransporation rate from an open field covered by grass, which grows actively and is not water stressed. Note that ET_o is equivalent to the potential evapotranspiration (i.e., $PET_{grass} = ET_o$), described in Chapter 1, for a grass surface. ET_o is expressed as millimeters of water per day (mm/day) for a given land area. The total volume of water transpired by a field of area A would be calculated by multiplying ET_o by A (where A is expressed in square millimeters if ET_o is in mm/day).

ET_o is highly sensitive to climatic conditions. Generalized or indicative values can be used when detailed climate information is not available readily (Table 5.4). K_c, like ET_o, depends on the climate. K_c also depends on the type of crop and its growth stage (Table 5.5). If the crop of interest transpires less water than grass, K_c will be less

Table 5.4 Indicative values of ET_0 for generalized climate zones. Data from Brouwer and Heibloem 1986.

Climate zone	Low daily temperature (<15°C)	Moderate daily temperature (15°C–25°C)	High daily temperature (>25°C)
Arid	4–6	7–8	9–10
Semiarid	4–5	6–7	8–9
Subhumid	3–4	5–6	7–8
Humid	1–2	3–4	5–6

Table 5.5 Minimum (top) and maximum (bottom) duration (days) of each growth stage for select crops. Data from Brouwer and Heibloem 1986.

Crop	Total	Initial stage	Crop development	Midseason stage	Late season stage
Barley/oats/wheat	120	15	25	50	30
	150	15	30	65	40
Bean	75	15	25	25	10
	90	20	30	30	10
Cabbage	120	20	25	60	15
	140	25	30	65	20
Cotton	180	30	50	55	45
	195	30	50	65	50
Maize, grain	125	20	35	40	30
	180	30	50	60	40
Melon	120	25	35	40	20
	160	30	45	65	20
Onion	70	25	30	10	5
	95	25	40	20	10
Pea	90	15	25	35	15
	100	20	30	35	15
Peanut	130	25	35	45	25
	140	30	40	45	25
Pepper	120	25	35	40	20
	210	30	40	110	30
Potato	105	25	30	30	20
	145	30	35	50	30
Sorghum	120	20	30	40	30
	130	20	35	45	30
Soybean	135	20	30	60	25
	150	20	30	70	30
Sugar beet	160	25	35	60	40
	230	45	65	80	40
Sunflower	125	20	35	45	25
	130	25	35	45	25
Tomato	135	30	40	40	25
	180	35	45	70	30

Table 5.6 Crop factors (K_c) for select crops' growth stages. Data from Brouwer and Heibloem 1986.

Crop	Initial stage	Crop development	Midseason stage	Late season stage
Barley/Oats/Wheat	0.35	0.75	1.15	0.45
Bean	0.35	0.70	1.10	0.90
Cabbage	0.45	0.75	1.05	0.90
Cotton	0.45	0.75	1.15	0.75
Maize	0.40	0.80	1.15	0.70
Melon	0.45	0.75	1.00	0.75
Onion	0.50	0.70	1.00	1.00
Pea	0.45	0.80	1.15	1.05
Peanut	0.45	0.75	1.05	0.70
Pepper	0.35	0.70	1.05	0.90
Potato	0.45	0.75	1.15	0.85
Sorghum	0.35	0.75	1.10	0.65
Soybean	0.35	0.75	1.10	0.60
Sugar beet	0.45	0.80	1.15	0.80
Sunflower	0.35	0.75	1.15	0.55
Tomato	0.45	0.75	1.15	0.80

Note: Values are for normal conditions; values should be adjusted for high and low humidity and wind speed.

than 1; if the crop of interest transpires more water than grass, K_c will be greater than 1. Indicative values are available for moderate climates (Table 5.6), and these values can be adjusted to account for high or low humidity or wind speed. High humidity and low wind speed suggest that less water will be evapotranspired than the indicative value for a moderate climate, so the K_c value can be reduced by about 0.05. Low humidity and high wind speed suggest that more water will be evapotranspired than the indicative value for a moderate climate, so the K_c value can be increased by about 0.05 (Brouwer & Heibloem, 1986).

To estimate the ET_{crop} over the entire growing period (Box 5.1), first we identify the crop, then indicate where the crop is to be planted, the planting date, and the ET_o values for each month (Step 1). Next we identify the duration of growth stages for our crop; we set up these stages with respect to the monthly growing season. We find the K_c value for each stage for the crop we are planting (Step 2). Once we have the K_c values for each stage, we can convert these values into monthly K_c values (Step 3) to calculate the ET_{crop} using the ET_{crop} equation (Step 4). ET_{crop} is expressed as millimeters of water per day (mm/day); to get the monthly values, we multiply ET_{crop} by the number of days per month (Step 4). All months are assumed to have 30 days for this analysis.

Box 5.1 Step-by-Step Approach to Calculating ET_{crop} (Millimeters/Month). Example adapted from Brouwer and Heibloem, 1986.

Given:
Crop: Tomatoes
Planting Date: 1 February
Days/month: 30
Climate: Arid (Tulare County, California)

Step 1

A. Based on Table 5.5, we know that tomatoes have a typical growing period between 135 and 180 days. If we go with 150 days and assume that each month has 30 days, our total growing period is 5 months: February to June.

B. Now that we know the months in our growing period, we can find the average daily temperature for the 5 months; this information is available online or through agricultural extension programs (Table B5.1.1).

C. Once we know the average daily temperature for each month in our growing season, we can look up indicative values of ET_o from Table 5.4.

Table B5.1.1 ET_o values for Step 1.

	February	March	April	May	June
Days per month	30	30	30	30	30
Temperature (°C)	17	19	23	28	32
ET_o (mm/day)	7	7	8	9	10

Step 2

A. We can use Table 5.4 to identify the days per stage for tomatoes.

B. The length of each stage can be shorter or longer than the 30 day months. Here we identify the dates for each stage (Table B5.1.2).

C. Each stage has an associated K_c value for each crop; here we use Table 5.6 to look up the K_c values (see Table B5.1.2).

Step 3

A. Here we convert each stage K_c values into monthly K_c values using weighted sums (Table B5.1.3).

1–5 March = Initial Stage ($K_c = 0.45$)
6–30 March = Crop Development Stage ($K_c = 0.75$)

$$\text{March } K_c = \left(\frac{5}{30} \times 0.45\right) + \left(\frac{25}{30} \times 0.75\right) = 0.70$$

Table B5.1.3 Conversion of growth stage K_c values to monthly K_c values for Step 3.

	February	March	April	May	June
Monthly K_c	0.45	0.70	0.95	1.15	0.85

Step 4

A. We use equation 1 to calculate ET_{crop} using the monthly ET_o and K_c values calculated above (Table B5.1.4). This value is mm/day.

B. To get the monthly $ET_{crop,}$ we multiple the daily value by 30 days.

C. If we compare the ET_{crop} over the entire growing period (1038mm/growing period) to indicative values of crop water needs in Table 5.7, we see that we are beyond the upper end of the range; this is not too surprising, because we are growing tomatoes in an arid climate. It is important to keep in mind that the indicative values show only a general range—climate factors can influence crop water need greatly.

Table B5.1.4 Final values for ET_{crop} by day and month for February through June for Step 4.

	February	March	April	May	June
ET_{crop} (mm/day)	3.2	4.9	7.6	10.4	8.5
ET_{crop} (mm/month)	96	147	228	312	255

Table B5.1.2 Dates and K_c values for Step 2.

Growth stage	Initial stage	Crop development	Midseason stage	Late season
Length	35 days	40 days	50 days	25 days
Dates	1 Feb to 5 Mar	6 Mar to 15 Apr	16 Apr to 5 Jun	6 Jun to 30 Jun
K_c	0.45	0.75	1.15	0.8

Table 5.7 Indicative ranges of crop water requirements. Data from Brouwer and Heibloem 1986.

Crop	Crop water required (mm/period)	Drought sensitivity[a]
Barley/oats/wheat	450–650	Low-medium
Bean	300–500	Medium-high
Cabbage	350–500	Medium-high
Cotton	700–1,300	Low
Maize	500–800	Medium-high
Melon	400–600	Medium-high
Onion	350–550	Medium-high
Peanut	500–700	Low-medium
Pea	350–500	Medium-high
Pepper	600–900	Medium-high
Potato	500–700	High
Sorghum	450–650	Low
Soybean	450–700	Low-medium
Sugar beet	550–750	Low-medium
Sunflower	600–1,000	Low-medium
Tomato	400–800	Medium-high

Note: These ranges provide a general guideline; more or less water may be required for the crop to produce high yields depending on the climate and soil.
[a] Low = drought resistant.

In short, the two important considerations for estimating **crop water need** are climate and crop type. The indicative ranges of crop water requirements provide generalized ranges, but more or less water may be required for the crop to produce high yields, depending on the climate (Table 5.7). Generally, some crops, like cotton, peppers, and sunflowers, require more water than others, such as cabbage and peas. Although it is not directly considered in this calculation, soil water holding capacity also plays a significant role in the requirement for irrigation.

5.7 Future Trade-offs

To understand future trade-offs facing agriculture and food, we first need to distinguish between the related, but different concepts of agricultural productivity and food security. Agricultural productivity focuses on crop and livestock harvesting. Food security focuses on access to sufficient, safe, and nutritious food to meet dietary needs for a productive life. The supply of food relies on physical and natural resources (i.e., land, water, and fertile soil); political and social structures (i.e., land tenure and labor; technology such as plows, water pumps, and fertilizers); and economic markets (FAO et al., 2017). Strong markets can increase access to the nutrients fundamental to healthy living and can increase resilience during natural catastrophes that destroy local agricultural production and periods of price volatility for staple foods.

Although there is a surplus of calories globally, millions of people are affected by chronic hunger or undernourishment each year. The regions that tend to use food imports to meet local demands are arid, land-limited, or both, but the success of this "import strategy" is positively linked to the region's economic strength (Porkka et al., 2017). Consequently, regions that tend to be food insecure are often the same regions where poverty rates are high. With projected population growth there is critical need for increased food production as well as a need for creating strong markets and connecting rural communities to trade networks.

The question of whether **organic agriculture** can produce enough food to feed the world and meet food security goals is complex and controversial. Some researchers argue that **organic agriculture** presents an opportunity, especially for rural communities without access to strong and reliable markets and trade. Because **organic agriculture** relies on the management of local resources, there may be a potential to create strong local agroecosystems. Strong agroecosystems often have a reduced reliance on external inputs such as synthetic fertilizers and have an increased reliance on crop diversity and livestock rotations, suggesting a lower chance of simultaneous yield failures. Nevertheless, **organic agriculture** and its practices vary regionally; as we have learned, certifications do not always focus on sustainable water resources management, which is critical to the future of agriculture.

The question of whether **conventional agriculture** can produce enough food to meet the world's sustainability goals is similarly complex and controversial. Although **conventional agriculture** boasts higher yields both on the plot and systems scale, it does it at the expense of environmental impacts.

The most appropriate agricultural and food system for one region may not be appropriate for another region—local context is key. The agricultural systems that work well in one region may work poorly in another region due to variations in local economic, social, and environmental contexts. Capital is a fundamental component for investments in machinery, fertilizers, pesticides, and irrigation infrastructure. Adoption of farming practices can be hindered by labor resources and cultural preferences. For example,

Box 5.2 Water Footprints for Food

We already have a general understanding of the water consumed for a number of vegetable and cereal crops via evapotransipration, but we have yet to explore how animal products compare. By measuring the amount of water that is consumed in the life cycle stages of different foods, we can gain a better understanding of which foods require small or large quantities of water. We call this measurement the water footprint. The water footprint includes the water used for the food product (i.e., green and blue water) as well as the water polluted during its production (i.e., gray water). Green water is rainwater. Blue water is surface water and groundwater applied by irrigation.

Generally speaking, water footprints are larger for animal products than crop products (Mekonnen & Hoekstra, 2010). Water footprints for animal products can be considered in terms of the volume of water per unit weight, per unit energy, per unit protein or per unit fat. The water footprint for cattle, in terms of volume per energy (kilocalorie), is ten times higher than for vegetables and twenty times higher than for cereals (Table B5.2.1). When we consider the volume of water per unit protein or fat, fruits and nuts rise above cattle (see Table B5.2.1).

Animals, such as cattle, sheep, pigs, and poultry, can graze natural landscapes for food, have supplementary diets with cultivated crops such as corn, alfalfa and hay, or have a mixture of grazing and supplementary diets. Animals that graze natural landscapes still consume water via their diet, because these landscapes are rainfed so the landscapes evapotranspire green water. Animals that have supplementary diets supported by irrigated crops will have a green water and blue water component to their water footprint. Approximately 87% of water for animal production is green water. Blue water represents 6%, and gray water represents 7% of the water footprint for animals (Mekonnen & Hoekstra, 2010). Animal diets vary by country, so water footprints for animals vary by country, just like water footprints for crops can vary by country.

To reduce water stress a number of options related to global food and agriculture have been proposed: reduce food loss and waste, increase the efficiency of food production through technological changes, and make dietary changes from meat-based diets to plant-based diets. Food production has significant impacts on the environment, and one way that we can assess that impact is to estimate if the planet is operating within its boundaries or if it is at risk of destabilizing ecosystem processes. Recent work suggests that, to keep the environmental impacts of food within all planetary boundaries, no single option will be enough; instead, a combination of the aforementioned options will be needed (Springmann et al., 2018).

Table B5.2.1 Water footprint for select food products. Data from Mekonnen and Hoekstra 2010. Presented with two significant figures only.

Crop	Liter per kilogram	Liter per kilocalorie	Liter per gram of protein	Liter per gram of fat
Vegetables	320	1.3	26	150
Fruits	960	2.1	180	350
Cereals	1,600	0.5	21	110
Nuts	9,100	3.6	140	47
Milk	1,000	1.8	31	33
Eggs	3,300	2.3	29	33
Meat				
Chicken	4,300	3.0	34	43
Pig	6,000	2.2	57	23
Sheep or goat	8,800	4.3	63	54
Cow	15,000	10.0	110	150

Note: The water footprint is presented in terms of water per unit weight, per unit energy, per unit protein, and per unit fat.

labor requirements can limit the uptake of no-till practices due to increases in weed growth or the shift of labor burdens among members of the household. In some communities, tasks like **tillage** may be performed typically by men, while tasks like weeding may be performed typically by women. If labor diversion is not reallocated in such circumstances, the burden of additional work may be placed on women (Giller et al., 2009).

Sustainable agricultural systems and sustainable food systems go hand in hand. There are many paths to sustainability for agriculture and food, and these paths will depend on how sustainability is defined and how we manage critical agriculture inputs such as water resources (Box 5.2). It is likely that we will need a combination of technological and social-cultural practices to promote a more sustainable and secure agricultural and food future.

5.8 Key Points

- There are many types of agriculture, and each type has a different emphasis. (Section 5.1.1)
- Farming practices that focus on healthy soil ecosystems increase the capacity of the soil to retain and release water optimally for crop growth. (Section 5.1.1)
- Conservation **tillage** and residues are beneficial in dry and hot climates. In wet and humid climates, conservation **tillage** and residues can lead to **waterlogging**. (Section 5.1.1)
- The three main types of irrigation (i.e., surface, sprinkler, and **microirrigation**) have pros and cons. (Section 5.2)
- **Irrigation efficiency** (*E*) represents the overall efficiency of the physical system, including water conveyance and water application. (Section 5.2)
- At the crop scale, crop yields using organic practices are, on average, about 67% to 81% the crop yields using conventional practices. Nevertheless, there is evidence that **organic agriculture** outperforms **conventional agriculture** during drought. (Section 5.3.1)
- The yield benefits of crop diversification and conservation **tillage** are more pronounced during dry conditions. (Section 5.3.1)
- Even though organic yields are lower than conventional yields, **organic agriculture** obtains a premium price for products,

which can make it more profitable per unit output than **conventional agriculture**. (Section 5.4.1)

- Although **microirrigation** has the highest investment costs of the three main types of irrigation, it is the most efficient irrigation system. (Section 5.4.2)
- Because **conventional agriculture** relies heavily on fertilizers and pesticides, and the manufacturing of these products is very energy intensive, indirect energy requirements are higher for **conventional agriculture** compared with **organic agriculture**. (Section 5.5)
- The amount of water required for crop growth depends highly on the temperature, humidity, wind speed, and solar radiation. Crop production in arid and windy climates requires more water than crop production in humid climates. (Section 5.6)
- The crop type and the stage of growth also affect the water required and the crop's vulnerability to water stress. (Section 5.6)
- Food security and agricultural productivity are interrelated but not the same. (Section 5.7)
- There are many paths to a more sustainable agricultural future and more food secure future. (Section 5.7; Box 5.2)

5.9 Example Problems

Problem 5.1. Identify three agricultural practices a conventional farm using **surface irrigation** could adopt to reduce **evapotranspiration** or increase soil moisture? Why would these practices reduce **evapotranspiration** or increase soil moisture?

Problem 5.2. What are the yield trade-offs between **conventional agriculture** and **conservation agriculture**?

Problem 5.3. Calculate the **irrigation efficiency** (E, %). The total water diverted to the farm from the source (V_d) is 10,500 liters. The volume of water that gets to the farm (V_f) is 10,000 liters. The volume of irrigation water stored in the root zone (V_s) is 9,500 liters. Based on the E_c, what type of conveyance infrastructure is being used? Based on the E_a, what type of irrigation system is being used?

Problem 5.4. Calculate and compare the ET_{crop} for cotton and peas grown (using the minimum period) in Tulare, California (arid conditions), with a February 1 planting date. How do the water requirements compare to the range of indicative values?

Problem 5.5. What inputs to the calculation in Problem 5.4 would change if you were to calculate and compare the ET_{crop} for cotton and peas grown in Dooley, Georgia, with a February 1 planting date? How would the water requirements change?

5.10 Suggested Reading

Seufert, V., & Ramankutty, N. (2017). Many shades of gray—The context-dependent performance of organic agriculture. *Science Advances, 3*(3), e1602638.

5.11 References

Amosson, S. H., Almas, L., Girase, J. R., Kenny, N., Guerrero, B., et al. (2011). Economics of irrigation systems. No. B-6113. College Station: Texas A&M AgriLife Communications. http://amarillo.tamu.edu/files/2011/10/Irrigation-Bulletin-FINAL-B6113.pdf

Brouwer, C., & Heibloem, M. (1986). *Irrigation water management.* Rome, Italy: Food and Agriculture Organization of the United Nations. http://www.fao.org/docrep/S2022E/s2022e00.htm#Contents

Brouwer, C., Prins, K., & Heibloem, M. (1989). Annex 1: Irrigation efficiencies. In *Irrigation water management: Irrigation scheduling.* Rome, Italy: Food and Agriculture Organization of the United Nations. http://www.fao.org/docrep/t7202e/t7202e00.htm#Contents

Connor, D. J. (2013). Organically grown crops do not a cropping system make and nor can organic agriculture nearly feed the world. *Field Crops Research, 144*, 145–147.

Crowder, D. W., & Reganold, J. P. (2015). Financial competitiveness of organic agriculture on a global scale. *Proceedings of the National Academy of Sciences, 112*(24), 7611–7616.

de Ponti, T., Rijk, B., & van Ittersum, M. K. (2012). The crop yield gap between organic and conventional agriculture. *Agricultural Systems, 108*, 1–9.

Fedoroff, N. V, Battisti, D. S., Beachy, R. N., Cooper, P. J. M., Fischhoff, D. A., et al. (2010). Radically rethinking agriculture for the 21st century. *Science, 327*(5967), 833–834.

Food and Agriculture Organization (FAO). (2016). AQUASTAT - Main Database. http://www.fao.org/nr/water/aquastat/data/query/index.html

Food and Agriculture Organization (FAO). (2017). *The state of food security and nutrition in the world.* Rome, Italy: Food and Agriculture Organization of the United Nations. http://www.fao.org/state-of-food-security-nutrition/en/

Gaudin, A. C. M., Tolhurst, T. N., Ker, A. P., Janovicek, K., Tortora, C., et al. (2015). Increasing crop diversity mitigates weather variations and improves yield stability. *PLOS One, 10*(2), e0113261.

Giller, K. E., Witter, E., Corbeels, M., & Tittonell, P. (2009). Conservation agriculture and smallholder farming in Africa: The heretics' view. *Field Crops Research, 114*(1), 23–34.

Grosso, S. J. Del, & Cavigelli, M. A. (2012). Climate stabilization wedges revisited: Can agricultural production and greenhouse-gas reduction goals be accomplished? *Frontiers in Ecology and the Environment, 10*(10), 571–578.

Irmak, S., Odhiambo, L. O., Kranz, W. L., & Eisenhauer, D. E. (2011). *Irrigation efficiency and uniformity, and crop water use efficiency.* Biological Systems Engineering: Papers and Publications 451. Lincoln: University of Nebraska. https://digitalcommons.unl.edu/biosysengfacpub/451

Kniss, A. R., Savage, S. D., & Jabbour, R. (2016). Commercial crop yields reveal strengths and weaknesses for organic agriculture in the United States. *PLOS One, 11*(8), e0161673.

Mekonnen, M. M., & Hoekstra, A. Y. (2010). *The green, blue, and gray water footprint of farm animals and animal products.* Value of Water Research Report Series No. 48. Delft, the Netherlands: UNESCO–IHE Institute for Water Education.

Peigné, J., Ball, B. C., Roger-Estrade, J., & David, C. (2007). Is conservation tillage suitable for organic farming? A review. *Soil Use and Management, 23*(2), 129–144.

Ponisio, L. C., M'Gonigle, L. K., Mace, K. C., Palomino, J., de Valpine, P., & Kremen, C. (2015). Diversification practices reduce organic to conventional yield gap. *Proceedings of the Royal Society B: Biological Sciences, 282*(1799), 20141396.

Porkka, M., Guillaume, J. H. A., Siebert, S., Schaphoff, S., & Kummu, M. (2017). The use of food imports to overcome local limits to growth. *Earth's Future, 5*(4), 393–407.

Rosset, P. M., & Altieri, M. A. (1997). Agroecology versus input substitution: A fundamental contradiction of sustainable agriculture. *Society & Natural Resources, 10*(3), 283–295.

Seufert, V., Ramankutty, N., & Foley, J. A. (2012). Comparing the yields of organic and conventional agriculture. *Nature, 485*(7397), 229–232.

Seufert, V., Ramankutty, N., & Mayerhofer, T. (2017). What is this thing called organic? How organic farming is codified in regulations. *Food Policy, 68*, 10–20.

Soil Science Glossary Terms Committee. (2008). Tillage. In *Glossary of Soil Science Terms 2008* (p. 92). Retrieved from https://www.soils.org/publications/soils-glossary#

Springmann, M., Clark, M., Mason-D'Croz, D., Wiebe, K., Bodirsky, B. L., et al. (2018). Options for keeping the food system within environmental limits. *Nature, 562*(7728), 519–525.

Tubiello, F. N., Rosenzweig, C., Goldberg, R. A., Jagtap, S., & Jones, J. W. (2002). Effects of climate change on US crop production: Simulation results using two different GCM scenarios. Part I: Wheat, potato, maize, and citrus. *Climate Research, 20*(3), 259–270.

CHAPTER 6

Energy Water Use

6.1 Introduction

Although agriculture is the largest demand-side sector for water withdrawals globally, thermoelectric power is the largest demand-side sector in the United States. This trend—large water withdrawals for energy resources—is often the case in industrialized countries that generate significant thermoelectric power via fossil fuels. Future water use will be affected in many ways by future energy choices. Not only are power plants fired by coal and natural gas large **greenhouse gas** emitters, they can be two of the largest water users. Decisions that surround the adoption and commercialization of our energy technology must consider reliability, economic costs, and environmental impact, in addition to water requirements. In short, like agricultural decisions, there are often a number of trade-offs that can both indirectly and directly affect water requirements.

To understand water for energy, we must first have a basic foundation in energy resources. Energy can be created from different sources (Table 6.1). These sources fit into two broad categories: **renewable energy** sources and **nonrenewable energy** sources. **Renewable energy** sources can be replenished continuously within human time scales. Examples include solar energy, geothermal energy, wind energy, biomass, and hydropower. Note that solar energy can be used to generate electricity using *photovoltaic cells (solar PV)* or by concentrating the sun's rays as a heat source (*solar thermal*). **Nonrenewable energy** sources, such as coal and natural gas, require special environmental conditions to form and take millions of years to develop. **Nonrenewable energy** reserves and **renewable energy** potentials vary spatially and, in some cases, temporally across different countries.

Renewable and **nonrenewable energy** sources are considered *primary energy sources*. The primary sources are converted to

Table 6.1 General information about energy resources examined within this chapter.

Name	Nonrenewable or renewable	Typical distribution method	Availability of primary energy source
Nuclear	Nonrenewable	Centralized: Power plant	Available in areas where uranium (^{235}U) is economically recoverable. Uranium is a metal that is naturally occurring in Earth's crust and seawater.
Petroleum (oil)	Nonrenewable	Centralized: Power plant	Available in areas where petroleum is economically recoverable. Petroleum is a mixture of hydrocarbons trapped within sedimentary rocks. Petroleum is formed from decomposed plants or animals under high pressures and heat over millions of years.
Coal	Nonrenewable	Centralized: Power plant	Available in areas where coal is economically recoverable. Coal is sedimentary rock high in hydrocarbons. Coal is formed from decomposed plants or animals under high pressures and heat over millions of years.
Natural gas	Nonrenewable	Centralized: Power plant	Available in areas where natural gas is economically recoverable. Natural gas is a mixture of hydrocarbons but primarily methane, trapped within sedimentary rocks. Natural gas is formed from decomposed plants or animals under high pressures and heat over millions of years.
Biomass	Renewable	Centralized: Power plant	Available in areas where residues are available or where land to grow dedicated feedstocks (i.e., poplar trees) is available.
Geothermal	Renewable	Decentralized: Household heat pumps Centralized: Power plant	Available in areas with geothermal heat potential. The potential is highest near tectonic plate boundaries.
Solar PV	Renewable	Decentralized: Household PV Centralized: PV farms	Available in areas with solar radiation potential, which varies spatially and temporally.
Solar thermal	Renewable	Centralized: Power plant	Available in areas with solar radiation potential, which varies spatially and temporally.
Wind	Renewable	Decentralized: Household, small turbines Centralized: Wind farm, large turbines	Available in areas with wind potential, which is dependent on solar radiation and land's ability to absorb heat.
Hydroelectric	Renewable	Centralized: Hydroelectric plant	Available in areas with moving water, dependent on precipitation, reservoir size, and balancing water needs for other reservoir and dam purposes.

useful energy (e.g., thermal energy through combustion or heating of water or steam) or kinetic energy (e.g., through using flowing water to turn a turbine). *Energy carriers* such as steam pipes or electrical transmission lines deliver energy services to end users. We can use primary energy sources to produce heat-based or thermal energy services, mechanical energy services, and electrical energy services (Moomaw et al., 2011).

Not all services provide the same benefits, and not all energy sources can provide the same services (Table 6.2). *Thermal energy*

Table 6.2 Services provided by energy resources examined within this chapter.

Name	Energy services		
	Thermal	Electrical	Mechanical
Nuclear	—	Fission is used to split uranium in a reactor with water, which releases heat (thermal conversion of energy). Heat is used to create steam, which turns a turbine (kinetic conversion of energy), powering a generator to generate electricity (energy carrier).	—
Petroleum (oil)	Direct use of thermal energy for heating	Fuels are burned (thermal conversion of energy) and converted to heat (energy carrier). Heat is used to create steam, which turns a turbine (kinetic conversion of energy), powering a generator to generate electricity (energy carrier).	Internal combustion of fuel (thermal conversion of energy). Combustion generates heat (energy carrier), which causes combustion gases to expand and initiates the rotation of gears (kinetic conversion of energy). Motion of gears drives rotation of wheels (energy carrier is mechanical work).
Coal	Direct use of thermal energy for heating	Fuels are burned (thermal conversion of energy) and converted to heat (energy carrier). Heat is used to create steam, which turns a turbine (kinetic conversion of energy), powering a generator to generate electricity (energy carrier).	—
Natural gas	Direct use of thermal energy for heating	Fuels are burned (thermal conversion of energy) and converted to heat (energy carrier). The heat is used to create steam, which turns a turbine (kinetic conversion of energy), powering a generator to generate electricity (energy carrier).	Internal combustion of fuel (thermal conversion of energy). Combustion generates heat (energy carrier), which causes combustion gases to expand and initiates the rotation of gears (kinetic conversion of energy). Motion of gears drives rotation of wheels (energy carrier is mechanical work).
Biomass	Direct use of thermal energy for heating	Biomass is burned (thermal conversion of energy) and converted to heat (energy carrier). Heat is used to create steam, which turns a turbine (kinetic conversion of energy), powering a generator to generate electricity (energy carrier).	Internal combustion of biofuel (thermal conversion of energy). Combustion generates heat (energy carrier), which causes combustion gases to expand and initiates the rotation of gears (kinetic conversion of energy). Motion of gears drives rotation of wheels (energy carrier is mechanical work).
Geothermal	Use of constant temperature in Earth's upper crust for heating and cooling	Heat or steam from deep groundwater wells (≤2 miles) is used directly. Heat (if applicable) is used to create steam, and the steam turns a turbine (kinetic conversion of energy) powering a generator to generate electricity (energy carrier).	—
Solar PV	—	Direct conversion of solar radiation to electricity (energy carrier).	—
Solar thermal	Direct use of thermal energy for heating	Sunlight, concentrated on reflective lenses, heats a fluid to create steam (thermal conversion of energy). The steam turns a turbine (kinetic conversion of energy), powering a generator to generate electricity (energy carrier).	—
Wind	—	Wind creates lift, turning blades (kinetic conversion of energy) to power a generator to generate electricity (energy carrier).	—
Hydroelectric	—	Gravitational force of falling water pushes a turbine (kinetic conversion of energy), powering a generator to generate electricity (energy carrier).	—

services are the provision of heat. For example, home heating is one type of thermal energy service. Primary energy sources like petroleum, natural gas, and biomass are converted into heat; heat becomes the energy carrier and provides thermal energy services. Many of our traditional modes of transportation—cars, planes, and boats—provide *mechanical energy services.* For example, gasoline and biofuels are converted into heat, which is then converted into mechanical work; mechanical work becomes the energy carrier and provides mechanical energy services. Electricity provides *electrical energy services*, which helps power devices and machinery that cannot be powered with primary energy sources. For example, our televisions and computers use electricity. Electricity is considered a *secondary energy source* because it is produced from primary energy sources that are converted into a usable form of energy. Secondary energy sources are neither renewable nor nonrenewable.

The world's *energy portfolio* is composed of both renewable and nonrenewable sources of primary energy as well as secondary

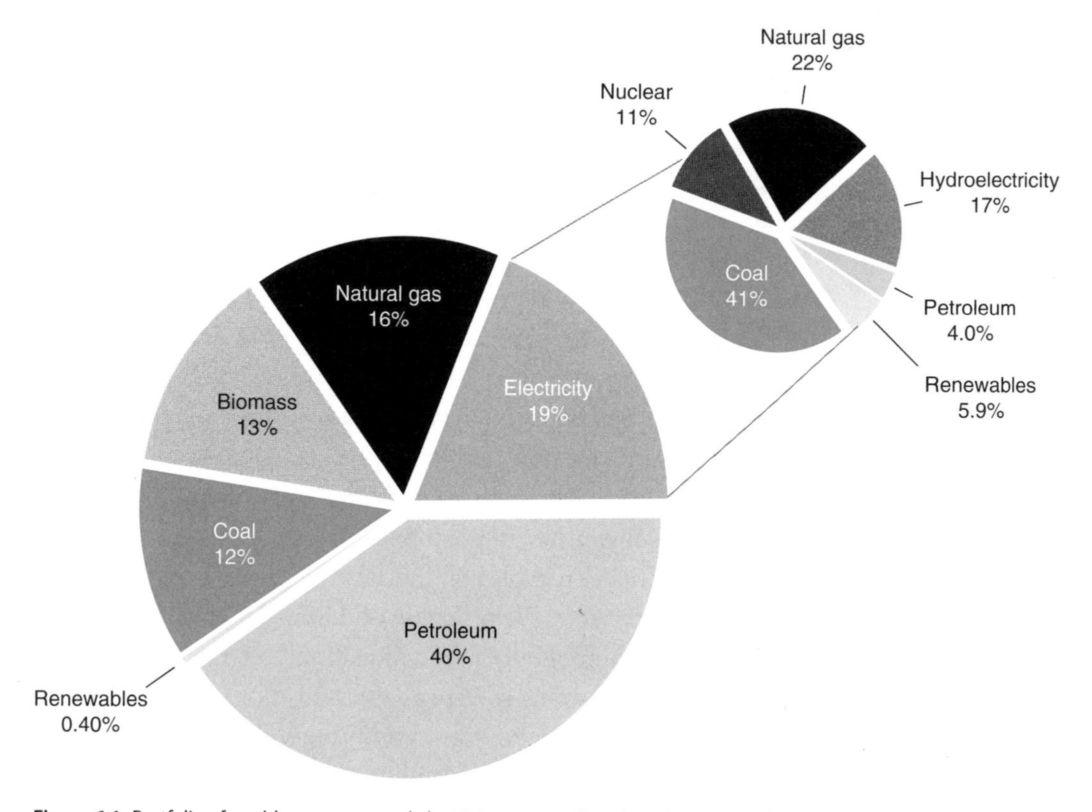

Figure 6.1 Portfolio of world energy use and electricity sources. Data from International Energy Agency 2014, 2016, 2017.

energy in the form of electricity (Figure 6.1). In this chapter we focus on electrical energy services and the eight primary energy sources that are likely to be used to generate electricity in the coming decades: nuclear, coal, natural gas, geothermal, solar, biomass, wind, and hydroelectricity (see Tables 6.1 and 6.2). Although oil is used currently in many countries to generate electricity, it is unlikely that future electricity growth will depend heavily on liquid petroleum products. Because many energy sources can produce electrical energy services, selecting an energy source requires an understanding of the energy technology, its reliability, its costs, its environmental impact, and its water demands.

6.2 Reliability

There are many facets to energy reliability in addition to adequate water resources. Here we will explore energy reliability through the availability and dependability of each of the eight energy resources likely to be used to generate electricity in the coming decades.

Energy *availability* is a function of the energy source itself. Is there enough natural gas or coal or any other energy source to meet the regional demand for electricity? If local sources of energy are not sufficient, energy availability depends on trade. **Nonrenewable energy** sources are traded on the international market, so even if a country does not have its own **nonrenewable energy** sources, it may have the ability to purchase **nonrenewable energy** sources on the international market. **Renewable energy** sources tend not to be traded directly on the international market, but the equipment to build power plants and the electricity generated from **renewable energy** sources can be traded among countries.

Energy *dependability* is a function of economic welfare and political stability in a region. Are there resources to build and maintain infrastructure so that energy services are provided uninterrupted and at affordable prices to the public? Dependable and affordable electricity provides reliable lighting, heating, cooking, and telecommunication services. Dependable and affordable electricity is a fundamental component of the acquisition, treatment, and distribution of water resources. Dependable and affordable energy provides important opportunities to promote economic growth.

The availability and dependability of electricity also relate to the capacity factor of the energy source and its associated electricity

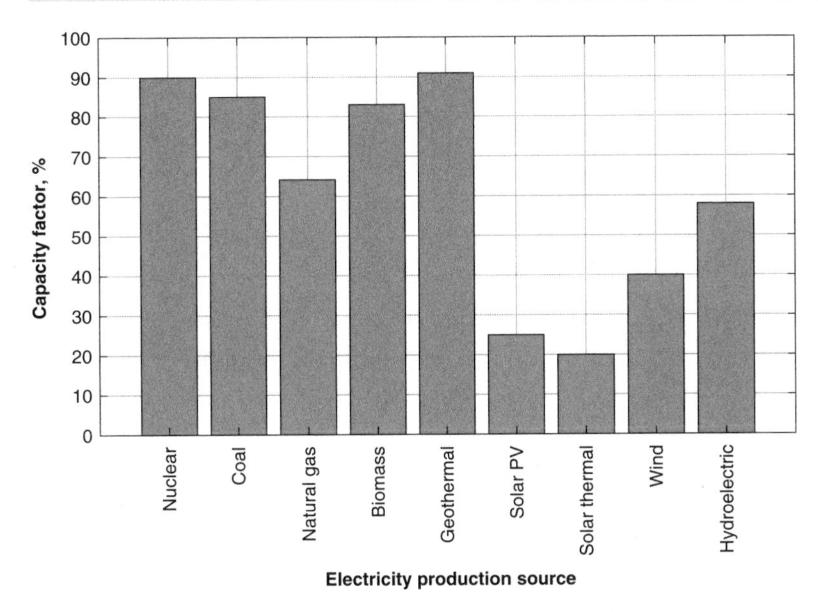

Figure 6.2 Average capacity factors for electricity production using nonrenewable sources (nuclear, coal, and natural gas) and renewable sources (biomass, geothermal, solar photovoltaic, solar thermal, wind, and hydroelectricity). Dispatchable technologies are flexible and can meet continual demand, so they have higher capacity factors. Nondispatchable technologies are less flexible, so they have lower capacity factors. Data from US Energy Information Administration 2017a, 2017b.

generation technology (Figure 6.2). The *capacity factor* is the ratio of actual electricity generation to maximum feasible electricity generation over a specific period of time (US Energy Information Administration, 2017a). **Nonrenewable energy** sources tend to have higher capacity factors than renewable sources, but there are some exceptions (see Figure 6.2). Higher capacity factors are often associated with technologies that are flexible and able to meet demand at any time, such as electricity generated in power plants using coal, natural gas, nuclear, geothermal, or biomass. Such technologies are referred to as *dispatchable* because power can be dispatched at the request of grid managers. Examples of *nondispatchable* energy sources include wind, solar PV, and solar thermal. Lower capacity factors are often associated with nondispatchable technologies because they are less flexible in following demand than nonrenewable sources. Hydroelectricity can be considered dispatchable within seasons when water is available via storage. Recent drought events may signal, however, that hydroelectricity is less dispatchable than previously considered, especially during prolonged dry seasons (US Energy Information Administration, 2017b).

Centralized distribution refers to an energy technology that generates electricity at one place and then distributes it from the point of generation through a network to end users. Moving electricity from one place to another requires an electricity transmission and distribution grid. The efficiency of moving electricity decreases as the distance between the point of generation and point of use increases because energy is lost in transmission. In the United States, transmission line loss represents about 5% of the electricity transmitted and distributed annually (US Energy Information Administration, 2017a).

Decentralized distribution refers to an energy technology that generates electricity at the same place as its end users (Moomaw et al., 2011). Decentralized energy systems can act as a stand-alone electricity source, or the systems can be connected to the electricity grid. If the decentralized energy system is connected to the grid, it has the added advantage of feeding surplus electricity into the grid and drawing electricity from the grid during shortages (Kaundinya et al., 2009).

Centralized distribution can be advantageous in areas where local energy resources may not be available to produce enough electricity to meet local demands. Decentralized distribution can be advantageous in very remote areas where the larger grid cannot be accessed. Because decentralized distribution can often generate electricity in closer proximity to local demands than centralized distribution, the transmission loss of electricity is reduced greatly. Electricity generated with **nonrenewable energy** sources tends to be centralized, whereas electricity generated with **renewable energy** sources has used both centralized and decentralized distribution setups (Alanne & Saari, 2006; Sathaye et al., 2011).

6.3 Economic Trade-offs

The cost of an energy technology plays a major role in its likelihood of adoption. Although actual costs can vary significantly depending on the region, demand, and existing electricity portfolio, we can look at average *levelized costs* to gain general insights into the costs for energy sources used to produce electricity. Electricity output is measured in *kilowatt-hours* (kWh), a unit equivalent to the energy needed to power 10 100-watt light bulbs for 1 hour;

Box 6.1 Life Cycle Assessments

A life cycle assessment is one technique used to quantify and evaluate the environmental costs and benefits of a product (ISO 14040:2006). Products can be goods (e.g., glass cups or plastic cups) or services (e.g., transportation via rail or barge). Life cycle assessments can be used to compare two or more products to quantify their relative environmental impacts. For example, we can compare different electrical services to make informed decisions about future development; these decisions may consider economic costs, emissions, and water use.

A life cycle assessment starts with the upstream phase, which includes the acquisition of raw materials and the manufacturing of equipment used to create a product. A life cycle assessment ends with the downstream phase, which includes all aspects of end-of-life treatment such as recycling and disposal. The middle phase incorporates the operation or use of the product. A life cycle assessment that looks at all these phases is often referred to as a "cradle-to-grave assessment" (ISO 14040:2006).

The upstream phase is sometimes referred to as the supply chain. The products or operations involved in these phases are not linked to direct operations of the product in scope; rather, they are part of another product series of production and distribution. The upstream phase for electrical services (Figure B6.1.1) can include two components: indirect and fuel cycle. The indirect phase incorporates the materials and equipment for the construction of a power plant. The fuel cycle phase is only applicable for some energy sources and includes the acquisition, processing, and transportation of the energy source.

Comparative life cycle analysis of electricity generation technologies indicates that electricity generation technologies that do not involve combustion of either fossil fuels or biomass have lower environmental impacts than the alternatives overall, even though the impacts for the indirect phase can be greater for solar and wind generation relative to fossil fuel technologies (Gibon, Arvesen, & Hertwich, 2017).

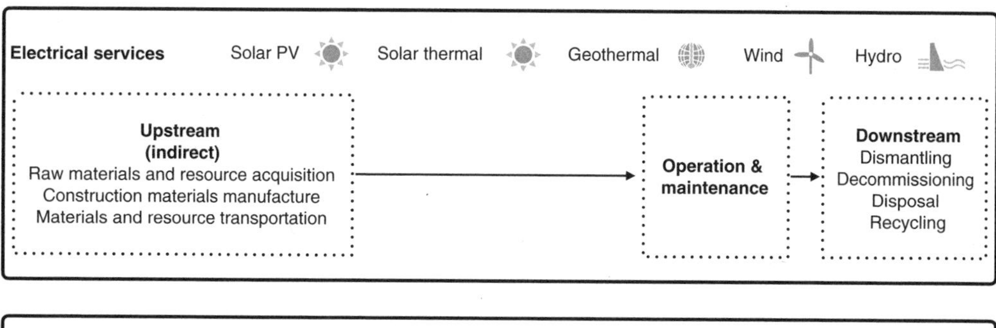

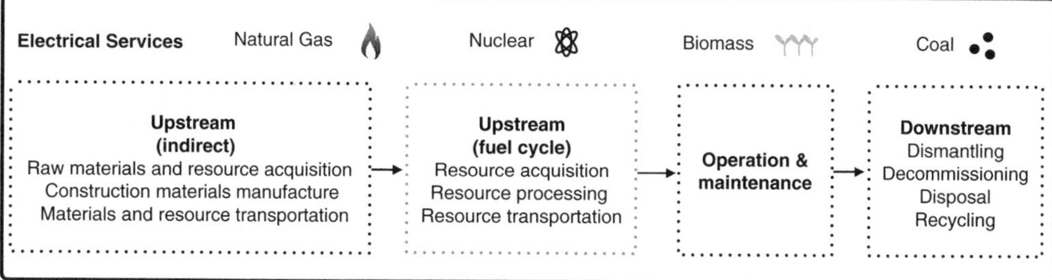

Figure B6.1.1 Life cycle phases for electricity generation as considered in Chapter 6.

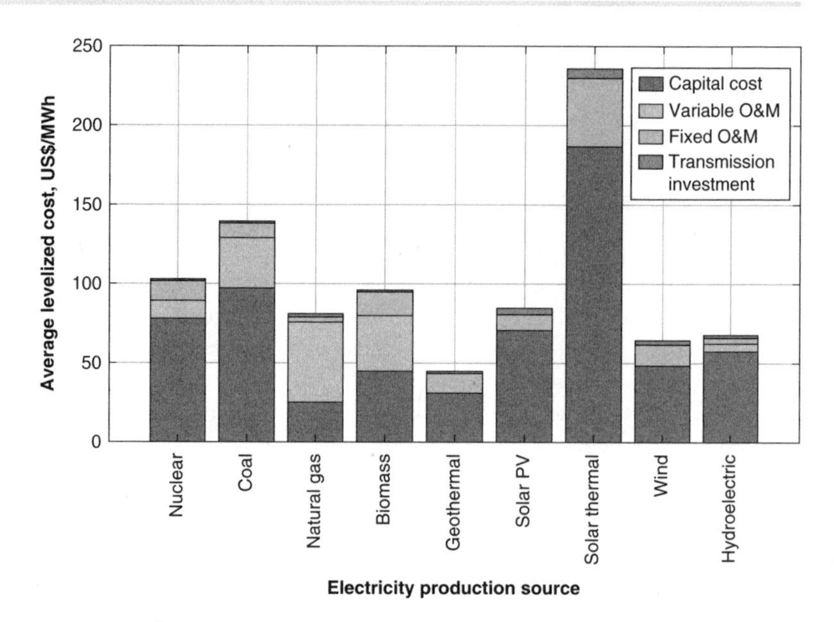

Figure 6.3 Average levelized cost of electricity for plants entering service in 2012. Upstream costs are for capital investment and for the fuel cycle (i.e., variable operation and maintenance [O&M] costs including fuel). Operations costs include fixed O&M costs and investment in transmission lines. Data from US Energy Information Administration 2017a.

1 megawatt is 1,000 kilowatts, so 10 100-watt light bulbs running for 1 hour uses 0.001 megawatts. Levelized costs are the costs per kilowatt-hour averaged over the life cycle of an energy source (Box 6.1).

The levelized cost metric can be useful in comparing the costs of construction and equipment (*capital costs*), *fuel costs* (if applicable), *operation and maintenance (O&M) costs*, and *transmission costs* for the eight energy sources examined in this chapter. We can think about costs in terms of life cycle stages, which compartmentalize the acquisition, processing, and use of energy resources. We can consider capital costs as part of the *upstream stage*, and fuel costs as part of the *fuel cycle stage*. We can consider transmission costs and O&M costs such as labor, site management, and corrective maintenance as part of the *operations stage*. Because most renewable resources do not have fuel costs, their levelized costs are primarily a function of capital costs. One inference from a levelized cost comparison for the United States is that most electricity plants using **renewable energy** sources are competitive economically with their nonrenewable counterparts (Figure 6.3).

6.4 Environmental Trade-offs

Life cycle analysis is one way that we can quantify the environmental impact of our energy resources. Life cycle analyses are useful because they allow us to quantify the impact at multiple stages of the energy resource. The raw materials acquisition and the environmental impacts of O&M and recycling can look very different depending on the energy sources, the life cycle stages of the chosen technology (see Box 6.1), and the countries or regions in which these stages take place. Here we compare the land requirements and the emission of greenhouse gases that affect climate change for each of the eight primary energy sources in terms of their *life cycle stages*.

All energy technologies require some amount of land throughout their life cycle (Box 6.2). Land requirements tend to be dominated by the fuel cycle for the energy sources that have a fuel cycle: coal, natural gas, biomass, and hydroelectricity (where we consider the water stored behind a dam to be the fuel). Land requirements are dominated by O&M for energy sources that do not have fuel cycles: solar PV, solar thermal, wind.

The distinction between life cycle stages is an important one, because not all land uses have the same environmental impact. Although all land uses disturb wildlife and their natural habitats, the land requirements for the upstream and fuel cycle stages of electricity production tend to be more destructive than the land required for the operations stage. Upstream land requirements are considered indirect land requirements: the land related to the materials and energy requirements for equipment and infrastructure.

The fuel cycle, which includes land for mining or processing of **nonrenewable energy** sources, dominates the life cycle land requirements for nonrenewable resources. Similarly, the fuel cycle for biomass, which includes land for cropping, dominates the life cycle land requirements. The fuel cycle stage often destroys natural habitat and vegetation; if it is not properly regulated, this stage has significant potential to contaminate the air, water, and soil. Development of large hydroelectric projects can inundate significant areas, displacing terrestrial ecosystems (Chapter 2 and Chapter 8).

Box 6.2 Land-Use Requirements

Land-use requirements can vary significantly across energy sources as well as across life cycle stages. Here we explore land use for nuclear, coal, natural gas, geothermal, solar PV, solar thermal, wind, and hydroelectric energy for each life cycle stage—indirect, fuel cycle, and operations and maintenance (O&M) (Figure B6.2.1).

For nuclear energy, the operations stage requires more land than other nonrenewable energy sources. Although the land needed for the power plant is similar to the land needed for coal and natural gas plants, nuclear plants require buffer areas for security purposes and to minimize the risk of exposure to radiation should an accident occur. As a result, a good portion of the land required for a nuclear power plant facility is not transformed significantly.

For coal, mining is especially destructive if surface mining is employed. One form of surface mining is referred to as *mountaintop removal* because the process often strips off the vegetation, topsoil, and rock above the coal seam and then deposits those materials in the valleys surrounding the mountaintop. The contours of the area change significantly.

The fuel cycle for natural gas, which includes drilling and extraction of the resource, requires the majority of its land use during its life cycle.

For geothermal energy, the power plant and its geothermal wells require the largest amount of land. Geothermal energy does not have a fuel cycle, and the upstream land requirements for materials are small compared with the other energy sources.

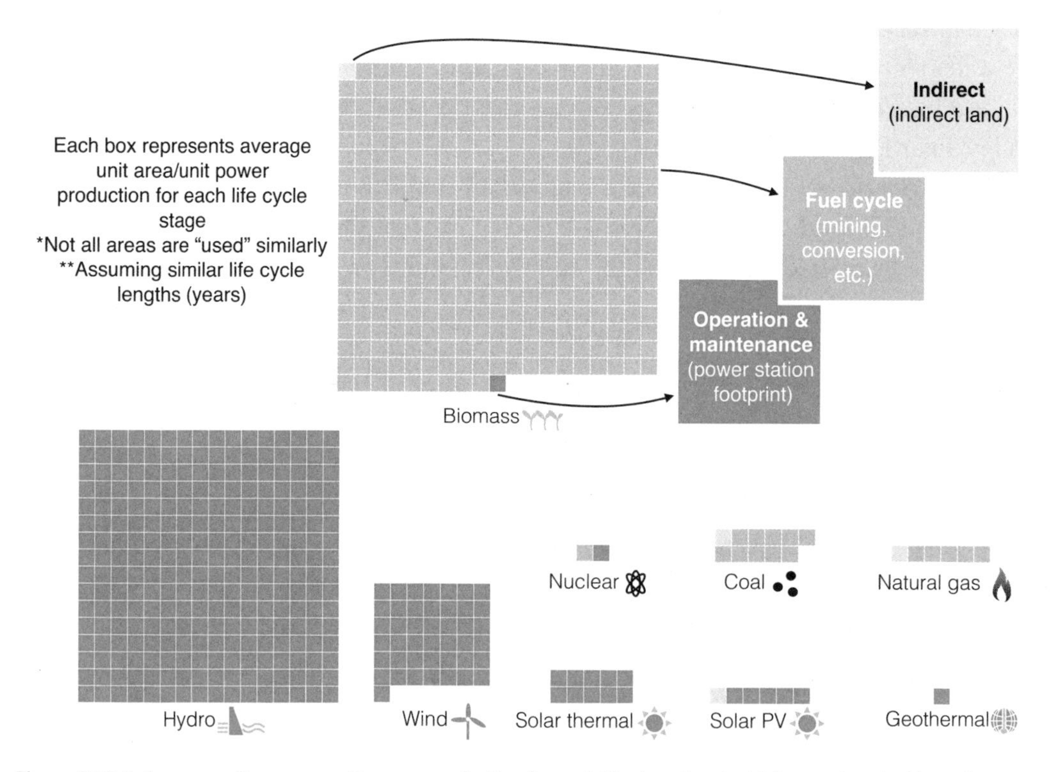

Figure B6.2.1 Average unit area per unit power production for each life stage for electricity produced with nuclear, coal, natural gas, geothermal, solar photovoltaic, solar thermal, wind, hydroelectricity (with reservoir), and biomass (cropland production) assuming 30-year life cycles. Land use is variable, depending on the specific energy technology. Data from Fthenakis and Kim 2009; Blodgett 2014.

Utility-scale PV farms require land for the configuration of PV panels. Rooftop PV, however, does not require designated land. The materials and energy that are required upstream to manufacture PV panels are not critical restrictions compared with the land required for a PV farm. Similarly, solar thermal power plants require land for the configuration of the mirrors that capture sunlight.

For wind energy, the land requirements are primarily for turbine spacing and configuration. Therefore, the land can be used for multiple purposes, such as for agricultural or recreation, in additional to wind farming.

Land use for hydroelectric facilities is associated with the reservoirs behind high-head dams. Dams have negatively impacted salmon habitat by restricting spawning grounds. Dams also reduce sediment flow downstream, which can negatively impact habitat and ecosystems downstream. Hydroelectricity can also be generated with run-of-the-river plants, which do not require large reservoir systems, greatly reducing their land requirements.

The land requirements for biomass vary widely. When land is devoted specifically to producing biomass for electricity generation, and the land is a direct conversion from natural habitat to cropland for biomass production, a significant amount of land is used. Most biomass sources have low energy efficiency, so when land is directly converted from natural habitat to cropland, biomass requires the largest amount of land to be transformed per unit power produced of all the energy sources reviewed in Chapter 6. In the United States, however, most electricity produced from biomass is from crop residues, wood waste, and methane emissions. The potential for such feedstocks is high in the Midwest and on the Pacific Coast. In cases where residues, wood waste, and methane emissions are used, the allocation of land transformation as a result of biomass use is more difficult to assess.

The land used for the operations stage looks different across energy sources. For **nonrenewable energy** sources, the land is used for the power plant. For most **renewable energy** sources, the land can often be used for multiple purposes, so "actual" land transformation is not easily assessed.

All energy technologies release some greenhouse gases throughout their life cycle. Greenhouse gases include carbon dioxide, methane, nitrous oxide, and fluorinated gases. The different gases stay in our atmosphere for different lengths of time, and they have different greenhouse effects. One common way to express the overall impact of emissions of all the greenhouse gases is through the use of a unit called *carbon dioxide equivalent* (CO_2e). The carbon dioxide equivalent unit converts the impact of all gases other than carbon dioxide to the impact of carbon dioxide. Combustible nonrenewable resources (e.g., coal, natural gas) release an order of magnitude more CO_2e than any other energy resources used for electrical services (Table 6.3).

Table 6.3 Carbon dioxide equivalent emissions (kg $CO_2e/MW{\cdot}h_{OUT}$) for electricity technologies by life cycle stage. Data from Burkhardt et al. 2012; Sullivan et al. 2013; Sullivan and Wang 2013; Turconi et al. 2013. Presented with two significant figures only.

Technology	Infrastructure (kg $CO_2e/MW{\cdot}h_{OUT}$)	Fuel cycle (kg $CO_2e/MW{\cdot}h_{OUT}$)	Plant operations and maintenance (kg $CO_2e/MW{\cdot}h_{OUT}$)
Coal	0 (0–1)	52 (16–84)	890 (770–970)
Natural gas	0 (0–1)	91 (42–130)	400 (380–410)
Oil	3	31 (19–38)	720 (630–820)
Nuclear	1.9 (1–3)	5 (3–13)	2 (1–3)
Biomass	0	24 (17–55)	3 (3–34)
Hydro	4 (4–7)	0	0
Solar PV	50 (39–90)	0	0
Solar thermal	16	0	170
Wind	15 (9–28)	0	0
Geothermal	6	0	0

Note: Medians presented with 25th and 75th percentiles in parentheses.

6.5 Water Trade-offs

Water is a critical requirement for the energy technologies that are used to generate electricity (Box 6.3). Water is used throughout the life cycle of all energy technologies (Tables 6.4 to 6.6). In the upstream life cycle stage, water is used for raw materials acquisition, for equipment, and for the construction of power plants. Depending on the energy technology, water may be used in the fuel cycle for mining and processing of the raw energy source. During the operations stage of electricity generation, water is used for cooling, if applicable, and cleaning.

In considering the water uses for electricity technologies we can distinguish between water withdrawals and water consumption. Water withdrawals include water removed from its source. For energy, water consumption is the portion of the water withdrawn that is lost through evaporation. The distinction between withdrawals and consumption is important because the portion of the water withdrawn, but not consumed, is eventually available for other downstream uses, but the portion of the water consumed is no longer available to other uses in the proximate area.

The withdrawal and consumption of water can vary significantly by energy source, technology, and life cycle stage (see Tables 6.4 to 6.6). Water withdrawal and consumption for the upstream stage are often negligible compared to the fuel cycle and operations stages. In fact, for the majority of energy sources discussed in this chapter, the water withdrawal and consumption requirements are domi-

Table 6.4 Water use for part of the upstream (indirect: power plant or technology construction) life cycle stage. Data from Bakken et al. 2013; Berndes 2008; Fthenakis and Kim 2009; Gerbens-Leenes et al. 2009; Lampert et al. 2015; Macknick et al. 2012; Meldrum et al. 2013; De Miranda Ribeiro and da Silva 2010; Solley et al. 1998. Presented with two significant figures only.

Technology	Withdrawal (m³/MW·h)	Consumption (m³/MW·h)
Nuclear	Minimal	Minimal
Coal	0 (0–0.1)	0 (0–0.1)
Natural gas	Minimal	Minimal
Biomass	Minimal	Minimal
Geothermal	Minimal	Minimal
Solar thermal	0.6 (0.4–0.6)	0.6 (0.3–0.6)
Solar PV	0.4 (0–6.1)	0.3 (0–0.8)
Wind	0.1 (0–0.3)	Minimal
Hydroelectricity	Minimal	Minimal

Note: Medians presented with minimum and maximum values in parentheses.

Table 6.5 Water use for part of the upstream (fuel cycle) life cycle stage. Data from Bakken et al. 2013; Berndes 2008; Fthenakis and Kim 2009; Gerbens-Leenes et al. 2009; Lampert et al. 2015; Macknick et al. 2012; Meldrum et al. 2013; De Miranda Ribeiro and da Silva 2010; Solley et al. 1998. Presented with two significant figures only.

Technology	Withdrawal (m³/MW·h)	Consumption (m³/MW·h)
Nuclear	0.4 (0–1.6)	0.3 (0–1.2)
Coal	0.3 (0–1.2)	0.1 (0–0.9)
Natural gas	0.3 (0–1.2)	0.3 (0–1.2)
Biomass	130 (0–422)	79 (0–260)
Geothermal	NA	NA
Solar thermal	NA	NA
Solar PV	NA	NA
Wind	NA	NA
Hydroelectricity	NA	NA

Note: Medians presented with minimum and maximum values in parentheses. Some technologies (e.g., geothermal, solar thermal, solar PV, wind, and hydroelectricity) do not have fuel cycles. NA, not applicable; PV, photovoltaic.

Table 6.6 Water use during the operations phase of electricity production. Data from Macknick et al. 2012; Meldrum et al. 2013. Presented with two significant figures only.

Technology	Withdrawal (m³/MW·h)	Consumption (m³/MW·h)
Thermoelectric		
Nuclear (once-through)	180 (87–230)	1.5 (0.4–1.5)
Nuclear (recirculating)	4.3 (3.0–9.8)	2.7 (2.2–3.4)
Coal (once-through)	110 (72–150)	0.5 (0.3–0.9)
Coal (recirculating)	2.5 (2.2–4.5)	2.0 (1.9–2.9)
Natural gas (once-through)	85 (90–110)	0.7 (0.4–1.2)
Natural gas (recirculating)	1.9 (1.9–2.9)	1.4
Biomass (once-through)	130 (76–190)	1.1
Biomass (recirculating)	3.3 (1.9–5.5)	2.1 (1.8–3.7)
Geothermal (recirculating)	0.1 (0–0.1)	0 (0–1.4)
Solar thermal (recirculating)	3.2 (3.0–2.5)	3.4 (2.1–7.2)
Nonthermoelectric		
Solar PV	0.1 (0–0.2)	0.1 (0–0.2)
Wind	Minimal	Minimal
Hydroelectricity	NA	17 (5.4–68)

Note: Data are medians of reported median, minimum, and maximum values linked to various energy and cooling technologies for each energy source. Medians are presented with minimum and maximum values in parentheses. NA, not applicable; PV, photovoltaic.

Box 6.3 How Much Water Does the Average American Use to Meet Monthly Electricity Needs?

To give context to water requirements for electricity, let us look at the average energy consumption for a residential customer in America. According to the US Energy Information Administration, the average American used about 910 kilowatt-hours of energy per month in 2014. Some of the highest operational water consumption factors for electricity generating technologies are around 4,000 liters per megawatt hour (e.g., coal fuel, upper range of recirculating cooling) and some of the lowest are in the double digits, about 20 liters per MWh (e.g., upper range of photovoltaic panels at the utility scale).

Let us use these values to estimate high and low values for water consumption. A megawatt hour is 1,000 KWh, so this means that each month the average American consumes about 3,640 liters of water just for their energy consumption alone if the energy is generated with coal fuel via cooling tower. This number is reduced significantly—to 18 liters—if the energy is generated with photovoltaic panels.

To put these numbers into perspective, we can compare them to the amount of water a bathtub holds. Let us say that the average bathtub holds about 230 liters of water. This means that the average person consumes about 16 bathtubs filled with water per month to satisfy their monthly electricity needs if the electricity is generated with coal fuel. If photovoltaic panels generate the electricity, only one-thirteenth of one bathtub of water is needed. Which energy sources we as a society pick can really impact our water resources demands.

nated by the operations stage. The two exceptions are solar PV and wind-generated electricity. For these energy sources, there is no fuel cycle, and more water is withdrawn and consumed during the upstream stage than during the operations stage.

6.5.1 Upstream Water Use

Water is withdrawn and consumed for the acquisition and manufacturing of construction materials for all power plant equipment and infrastructure (see Table 6.4). Of the eight energy sources we explore in this chapter, water withdrawal and consumption during the upstream stage is highest for solar PV and solar thermal.

6.5.2 Fuel Cycle Water Use

For energy sources with a fuel cycle, water is withdrawn and consumed throughout the fuel cycle stage (see Table 6.5). Uranium (nuclear energy), coal, and natural gas require water for mining, as well as water for cleaning and processing raw energy materials into usable energy resources. The amount of water required varies with mining practices, purity of the resource, and processing approach. In the use of biomass for electricity generation, water is required for the growth of vegetation. The amount of water withdrawn and consumed depends highly on the vegetation grown and whether the crops are rainfed or irrigated. When the crops are irrigated, biomass rises as the energy source with the highest water withdrawal and water consumption, with water withdrawals about twice the water consumed. These high withdrawal values may indicate irrigation inefficiencies, a topic we discuss in Chapter 5.

6.5.3 Operations Water Use

Water withdrawal and consumption during the operations stage of electricity generation are linked to the energy technology, with the main distinction being between **thermoelectric** or nonthermoelectric technologies. **Thermoelectric** technologies use heat to generate electricity by (1) burning an energy source (thermal conversion of energy), converting the source to heat (energy carrier) (coal, natural gas, biomass); (2) using fission to split uranium to release heat (nuclear); or (3) using natural heat sources, such as heat from Earth's crust (geothermal) or heat from the sun (solar thermal). The heat is used to convert a liquid into a gas, which turns a turbine (kinetic conversion of energy) to power a generator to generate electricity (energy carrier) (e.g., see Table 6.2). When the liquid is water, it is referred to as *boiler water*, and the gas is referred to as *boiler steam*. After electricity is generated, the boiler steam is cooled, condensing back into boiler water. Traditionally, cold water from rivers is piped into a **thermoelectric** power plant or station to cool the gas; this cold water is referred to as *cooling water*. Cooling water is distinct from boiler water.

It is the cooling water that often dominates the total life cycle water withdrawals and water consumption for a **thermoelectric** power system (see Table 6.6). The amount of water required for the cooling process is primarily a function of (1) the thermal efficiency of the source of heat, (2) the turbine process(es), and (3) the cooling system. Generally, **thermoelectric** power systems that burn an energy source (a fossil fuel or biomass) can use a single cycle or a combined cycle. A *single cycle* uses a steam turbine process, where water is converted into steam to turn the turbine. A *combined cycle* uses both a steam turbine process and a gas turbine process. The gas turbine process uses natural gas or coal that has been converted into syngas to directly power turbines. In the combined cycle system, heat is recovered from the gas turbine process and used in the steam turbine process; as a result, combined systems are very efficient at creating electricity (Feeley et al., 2008). **Thermoelectric** power systems that use fission or natural heat sources (e.g., geothermal) use steam turbines without the combustion of fuels. The more efficient the turbine process at creating electricity, the less cooling water is required per unit electricity.

In thermoelectric power generation, the type of cooling system is a fundamental component to determining the total amount of water withdrawn and consumed during the operations phase (see Table 6.6). The cooling system is responsible for converting the boiler steam used to turn the turbine back into boiler water. Both air and water can be used for cooling the boiler steam. The colder the cooling air or water, the more efficiently the boiler steam condenses back into boiler water.

For **thermoelectric** technologies that burn an energy source or use fission, the three types of cooling systems are **dry cooling**, **once-through cooling**, and **recirculating cooling**. Water withdrawals for **once-through cooling** systems are orders of magnitude greater than for **recirculating** and **dry cooling** technologies. Water consumption for **recirculating cooling** technologies is nearly twice the water consumption for once-through technologies. Generally, **dry cooling** systems withdraw and consume the least amount of water for thermoelectric operations. Nevertheless, there are variations in cooling water requirements even within cooling technologies (Figure 6.4) because of different power plant technologies, energy sources, and climates.

Of the three cooling systems, the electricity generated with **once-through cooling** is the least expensive and most efficient (Figure 6.5). Compared with once-through systems, **dry cooling** systems incur a 4.5% energy loss (Tidwell et al., 2014), and **recirculating cooling** systems incur a 1% energy loss (DeNooyer et al., 2016). **Dry cooling** significantly reduces the life cycle water requirements of the operations stage because air is used to cool the boiler water. For **thermoelectric** power systems that use **once-through** or **recirculating cooling** systems, the total life cycle water requirements are dominated by the operations stage.

Once-through cooling systems withdraw large amounts of water, run the water through tubes in the condenser, and then discharge the water back to its source. Within this system, a lot of water is

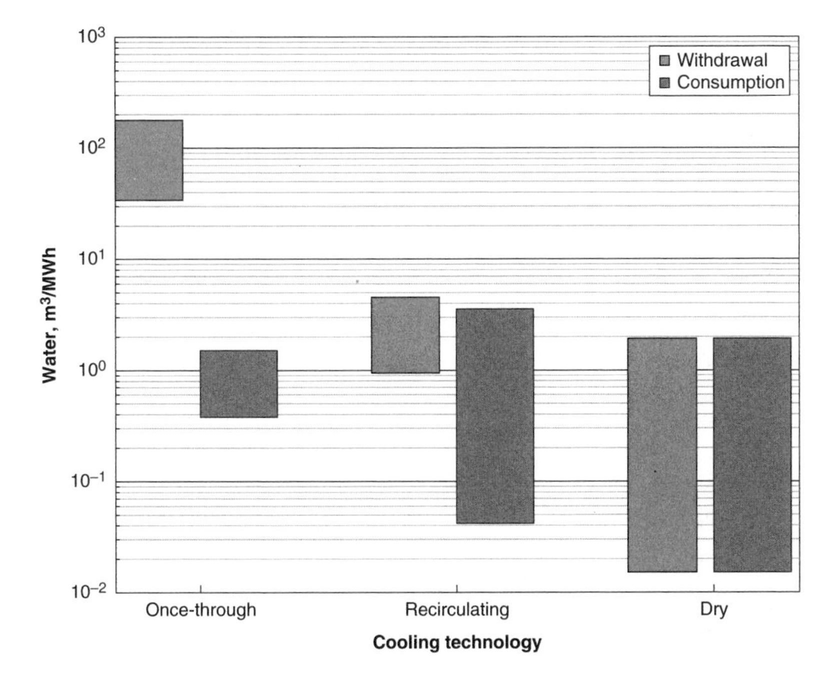

Figure 6.4 Generalized trends for cooling water withdrawal and consumption for cooling technologies at thermoelectric power plants. Note the logarithmic scale—the withdrawals for once-through cooling are at least 10 times larger than withdrawals for other technologies. Data from Macknick et al. 2012; Meldrum et al. 2013.

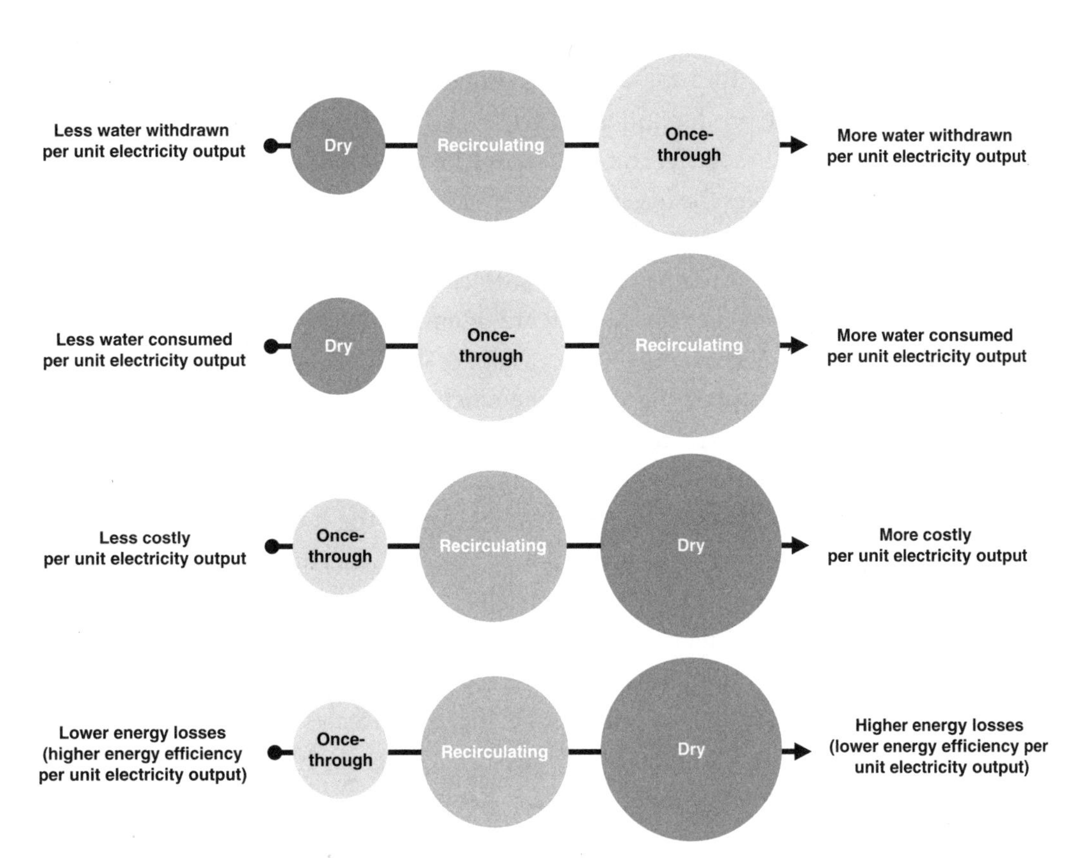

Figure 6.5 Water requirements, cost, and efficiency trade-offs for three types of cooling systems.

withdrawn, and only a little water is lost to evaporation (i.e., consumed). The water that is not lost to evaporation is discharged back to the source warmer than when it first entered the condenser.

Like **once-through cooling** systems, **recirculating cooling** systems withdraw water and run it through tubes in the condenser. Unlike once-through systems, recirculating systems use cooling towers or cooling ponds to dissipate heat from the warmed, cooling water to the atmosphere so that the cooling water can be recirculated through the tubes in the condenser again. As heat dissipates from the warmed cooling water to the atmosphere, some of the cooling water evaporates. Consequently, **recirculating cooling** systems withdraw less water than once-through systems, but recirculating systems consume more water than once-through systems (Table 6.6). We emphasize that cooling water is maintained separately from the boiler water. In fact, the mass flow rate of cooling water is more than 50 times the mass flow rate of boiler water (Feeley et al., 2008).

The cooling efficiency of **dry**, **once-through**, and **recirculating cooling** systems relies heavily on cool air or cool water. Consequently, heat waves and droughts can have significant impacts on the production of electricity. An increase in temperature will have a direct effect on **dry cooling** because the air will be warmer, which makes the dissipation of heat from the cooling water to the air less effective. An increase in temperature also will have a direct effect on the temperatures of surface water. **Thermoelectric** systems that use water for cooling are often built along major river corridors to access cool, flowing water easily. If the temperature of the cooling water increases, the efficiency of electricity generation will decrease. An increase in temperatures during winter and spring months also will accelerate snowpack melt, cause earlier runoff, and thus reduce river flows during the hottest times of the year. Without a sufficient flow in the river, **thermoelectric** power plants must curtail their electricity production.

Production curtailments during droughts and heat waves are common. Between 2000 and 2015, about 40 incidents of curtailment were documented in the United States as a result of high water temperatures or low flows (McCall et al., 2016). It is likely that the number of documented curtailments will increase as **anthropogenic** climate change continues to increase temperatures (Chapter 9). Recent research has suggested that **anthropogenic**-induced climatic changes will likely decrease the capacity of thermoelectric power production in the United States and Europe during summers by about 10% (Van Vliet et al., 2012).

Thermoelectric systems using **once-through cooling** have the potential of resulting in **thermal pollution** of rivers because the water discharged back to its source is warmer than when it first entered the condenser. **Thermal pollution** from the discharge of cooling water can increase river temperatures as much as 12°C (McCall et al., 2016). Consequently, **thermoelectric** system discharges can be detrimental to aquatic habitats during the summer. During the summer, river flows are lowest and solar radiation is highest, so river water is often at its warmest, regardless of **thermoelectric** system discharges. Higher-than-natural water temperatures can reduce the amount of dissolved oxygen in the water, which impacts nutrient cycling and habitat health, inhibits the growth and reproduction of aquatic animals, and increases the vulnerability of aquatic animals to toxic chemicals (Chapter 12).

For **thermoelectric** technologies that use natural heat sources (geothermal or solar thermal), the cooling systems are similar to those already explained. *Geothermal electricity* is generated by using steam or hot water heated within the Earth. A common technology is the binary system, in which the natural hot water, which can have corrosive chemical properties, is used to heat another liquid to produce steam. In the United States, most geothermal binary systems are moving to **dry cooling** (Clark et al., 2010). In geothermal systems, fresh water is withdrawn and consumed during the operations phase to minimize corrosion, manage the buildup of dissolved solids, or replenish the geothermal reservoir.

Solar thermal electricity is generated by concentrating solar radiation on a reflective lens to heat a fluid to create steam. Most solar thermal electricity plants operate trough systems, which use curved mirrors to focus solar radiation to heat water and create steam. To condense the steam back into water, trough systems can use **dry cooling, recirculating cooling,** or a hybrid system that uses both **dry** and **recirculating cooling**.

For nonthermoelectric energy technologies (e.g., solar PV, wind, and hydroelectricity), water withdrawals and consumption during the operations stage of electricity generation energy are not associated with cooling technology. During the operations stage of *solar*

PV and *wind*, minimal water is required. Water is used to occasionally wash PV panels and wind turbine blades. For *hydroelectricity*, water withdrawals are irrelevant because water is not removed from its source; hydroelectricity is generated by the gravitational force of falling water as it pushes a turbine (kinetic conversion of energy) to power a generator.

Water consumption during the operational stage of hydroelectricity depends on evaporative losses. The evaporative losses depend on the facilities used to generate electricity. There are three types of facilities used to generate hydroelectricity: (1) *multiuse reservoirs*, where water is stored and released to meet recreational, flood protection, water supply, or hydroelectricity demands; (2) *dedicated reservoirs*, where water is stored and released for hydroelectricity demands only; and (3) *run-of-the-river*, where little to no water is stored. The larger the reservoir's surface area, the greater the storage and the greater the evaporative flux (i.e., consumed water). Reservoirs can use naturally existing water bodies, such as lakes, to enhance storage. Reservoirs can also be built on ephemeral rivers, so the water can be artificially stored during periods of high precipitation and released during periods of low precipitation.

Determining water consumption of hydroelectricity production is complicated, because reservoirs can be used for many purposes. When a reservoir is dedicated to hydroelectricity production, the reservoir storage allows for the regulation of flows specifically with hydroelectricity production in mind, so the increased water evaporation from the impoundment over that from the natural system before the construction of the dam is attributable to power generation. When a reservoir is used for multiple purposes, hydroelectricity is not the single focus and may not even be the primary focus. Consequently, the allocation of evaporative losses among the many reservoir uses can be controversial: is the water allocated by the primary purpose of the reservoir's construction, by economic value attributed to each purpose, or equally among all purposes?

Unlike reservoir facilities, run-of-the-river facilities are located on rivers with traditionally consistent flows year-round. Therefore, these facilities require little to no storage to maintain consistent electricity production. As a result, run-of-the-river facilities tend to not increase the surface area of water exposed to solar radiation. Because the water consumption for run-of-the-river facilities is similar to the water system before the construction of the hydro-electricity facility, the operational water consumption is considered to be negligible.

6.6 Future Trade-offs

Future cost, environmental impacts, and water use of electricity generation vary greatly depending on the electricity generation technologies and the electricity generation scenarios that are realized. As we learned, nonrenewable sources of energy that are combustible contribute more to **greenhouse gas** emissions than nuclear and renewable sources of energy. To combat the high emissions associated with coal and natural gas, there are a number of paths forward. For example, we can use carbon capture and storage technologies, transition to other sources of energy for electricity generation, or increase conservation measures to limit the demand for electricity. Choices of future paths must consider the trade-offs, including water withdrawal and consumption.

The costs for the newer electricity generation technologies are higher than the cost for technologies that have been used for decades. This is especially true for many renewable electricity technologies. However, most renewable technologies do not have fuel stock costs, which suggests that their long-term costs may be more stable than the costs of **nonrenewable energy** on the global market. As we invest in research to increase the maturity of renewable technologies and allow for renewable technologies to penetrate the market, it is likely that costs will go down.

The goal of *carbon capture and storage (CCS)* technologies is to capture the carbon dioxide released at coal and natural gas power plants and to sequester it so that it is not released into the atmosphere. Although CCS could be effective in reducing **greenhouse gas** emissions, it reduces the thermal efficiency of the electricity generation and increases the processing requirements for coal and natural gas. As a result, CCS can increase water withdrawals between 60% and 100% and water consumption between 70% and 80% during operations (Meldrum et al., 2013). The fuel cycle water requirements and power plant water requirements also increase when CCS is used. The barriers to implementation of large-scale CCS are challenging, so it is not likely that CCS will see large-scale adoption in the immediate future.

The transition from coal and natural gas to nuclear or renewable sources of electricity can also pose water requirement trade-offs. The operations phase of nuclear energy requires a significant amount of water and, depending on the cooling system employed, can have large water requirements. Although many **renewable energy** sources are less water intensive, solar thermal typically has large water use requirements as well.

To combat the high freshwater use requirements for cooling systems associated with thermoelectric power, cooling system retrofits from **once-through** and **recirculating cooling** with fresh water to **recirculating cooling** with municipal wastewater and brackish water or **dry cooling** are being considered in a number of locations. Current levelized cost estimates suggest that, on average, retrofits to **recirculating cooling** systems with wastewater are the most cost-effective transitions (Tidwell et al., 2014). Retrofits to **dry cooling** are most costly, and these retrofits are considered mostly where wastewater and brackish water are unavailable.

In the western United States, water rights can restrict the use of wastewater for new uses if the water is claimed downstream by users who plan to treat the water. Characterizing brackish water availability is difficult due to the complexity of geology and aquifers as well as the patchwork of state water rights that govern **groundwater**. Although all three types of retrofits increase the levelized costs of electricity, in most areas, the increased cost per unit electricity would be less than 10% of current costs (Tidwell et al., 2014). Furthermore, all three types of retrofits would reduce water withdrawal and consumption significantly and increase resilience against droughts. Nevertheless, moving away from using fresh water and **once-through cooling** also comes with an energy penalty or reduced energy efficiency.

Water is a local, regional, and global resource, so it is also important to consider withdrawal and consumption at many levels when making decisions. That is, one set of electricity technologies may lower national water withdrawal and consumption, but also increase local water withdrawal and consumption in some places. Technologies that work well in arid regions may not work well in humid regions, and what works well today, may not meet the needs of the future. Although the selection of energy technologies raises a number of trade-offs and considerations, we do have high confidence that increased electricity efficiency and conservation will result in the reduction of electricity use, and consequently in reduction of water withdrawal and consumption.

6.7 Key Points

- Energy can be created from **renewable energy** sources and **nonrenewable energy** sources to produce heat-based or thermal energy services, mechanical energy services, or electrical energy services. (Section 6.1)
- Electricity is a secondary source of energy because it is produced from primary energy sources that are converted into a usable form of energy. (Section 6.1)
- Most electricity plants using **renewable energy** sources are competitive economically with their nonrenewable counterparts. As the technology for **renewable energy** matures, the costs are likely to go down. (Section 6.3)
- Life cycle assessments are useful evaluation techniques because we can explore environmental impacts throughout a product's life cycle. (Section 6.4, Box 6.1)
- The amount of land required and the greenhouse gases released for each energy technology vary greatly. (Section 6.4, Box 6.2)
- All energy technologies require some quantity of water over their life cycle, but the quantity varies by energy technology and life cycle stage. (Section 6.5)
- Significant water requirements are needed for cooling. (Section 6.5.3)
- One of the most effective ways to reduce environmental impacts, costs, and water use is to reduce the total demand for electricity. (Section 6.6)

6.8 Example Problems

Problem 6.1. An electricity portfolio is the relative (%) and absolute (#) value of each energy resource that provides primary energy to produce electricity. This information can be presented in a pie chart to show both the relative (%) values for each type of energy as well as the absolute (#) values for each type of energy. For example, California's electricity portfolio for 2017 looks like Figure 6.6.

a. For each named energy source in California's electricity portfolio, use the emissions factors in Table 6.3 to calculate yearly CO_2e. What are the per capita CO_2e for California's electricity portfolio? Use the

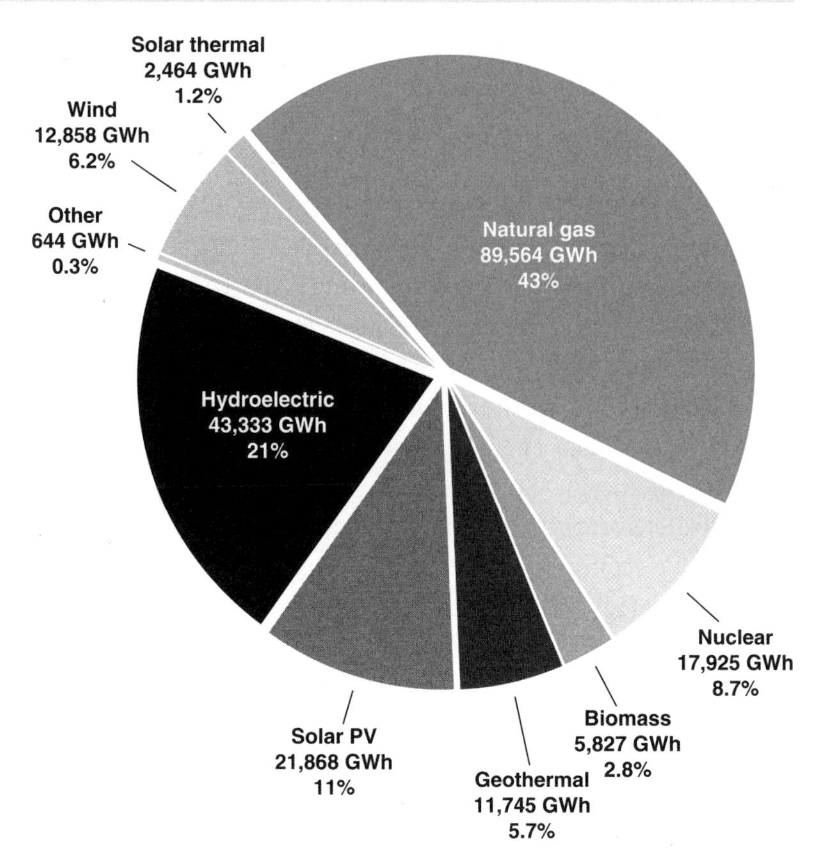

Figure 6.6 California's electricity portfolio for 2017. The "other" category includes coal, oil, petroleum coke, and waste heat. Data from California Energy Commission.

median value for CO_2e in Table 6.3. *Hint:* The population in 2017 was 39.5 million people, and 1 gigawatt hour (GWh) = 1,000 megawatt hours (MWh).

b. Use the water withdrawal factors in Tables 6.4 to 6.6 to calculate the yearly water withdrawals for California's energy portfolio assuming all **recirculating cooling** for **thermoelectric** power plants. What are the per capita water withdrawals for California's electricity portfolio? Use the median value for water withdrawal in Tables 6.4 to 6.6.

c. In 2018, California passed a bill to shift its electricity to 100% renewables by 2045. The state has incentivized wind and solar investments, suggesting that wind and solar are likely to replace the state's electricity generated from natural gas. Generally, utility-scale solar plants use solar thermal technology. How would switching from natural gas to solar thermal impact CO_2e emissions and water withdrawals?

Problem 6.2. Discuss generalizable trade-offs between **renewable** and **nonrenewable energy** sources, focusing on (1) reliability and dependability, (2) economic, (3) environmental, and (4) water trade-offs. Is there one clear energy technology for the area where you live? Why or why not?

Problem 6.3. It is the middle of the summer, and the local river is experiencing low flows because of an ongoing drought. Discuss the impact of once-through and recirculating power plants on the river's water quantity and quality.

6.9 Suggested Reading

US Environmental Protection Agency. (2017). Energy and the environment: Power profiler. https://www.epa.gov/energy/power-profiler

6.10 References

Alanne, K., & Saari, A. (2006). Distributed energy generation and sustainable development. *Renewable and Sustainable Energy Reviews, 10*(6), 539–558.

Bakken, T. H., et al. (2013). Water consumption from hydropower plants—review of published estimates and an assessment of the concept. *Hydrology and Earth System Sciences, 17*(10), 3983–4000.

Berndes, G. (2008). Future biomass energy supply: The consumptive water use perspective. *International Journal of Water Resources Development, 24*(2), 235–245.

Blodgett, L. (2014). Geothermal basics—environmental benefits. Geothermal Energy Association. http://geo-energy.org/geo_basics_environment.aspx

Burkhardt, J. J., Heath, G., & Cohen, E. (2012). Life cycle greenhouse gas emissions of trough and tower concentrating solar power electricity generation. *Journal of Industrial Ecology, 16*, S93–S109.

Clark, C. E., Harto, C. B., Sullivan, J. L., & Wang, M. Q. (2010). *Water use in the development and operation of geothermal power plants.* Argonne National Laboratory Report ANL/EVS/R-10/5. Oak Ridge, TN: US Department of Energy. https://www.energy.gov/sites/prod/files/2014/02/f7/geothermal_water_use_draft.pdf

De Miranda Ribeiro, F., & da Silva, G. A. (2010). Life-cycle inventory for hydroelectric generation: a Brazilian case study. *Journal of Cleaner Production, 18*(1), 44–54.

DeNooyer, T. A., Peschel, J. M., Zhang, Z., & Stillwell, A. S. (2016). Integrating water resources and power generation: the energy—water nexus in Illinois. *Applied Energy, 162*, 363–371.

Feeley, T. J., Pletcher, S., Carney, B., & McNemar, A. T. (2008). Department of Energy, National Energy Technology Laboratory, Power Plant–Water R&D Program. Presented at the Water- Energy Nexus Conference in December 2008. https://www.researchgate.net/publication/245866647_Department_of_EnergyNational_Energy_Technology_Laboratory's_Power_Plant-Water_RD_Program

Fthenakis, V., & Kim, H. C. (2009). Land use and electricity generation: A life-cycle analysis. *Renewable and Sustainable Energy Reviews, 13*(6–7), 1465–1474.

Gerbens-Leenes, P. W., Hoekstra, A. Y., & van der Meer, T. (2009). The water footprint of energy from biomass: A quantitative assessment and consequences of an increasing share of bio-energy in energy supply. *Ecological Economics, 68*(4), 1052–1060.

Gibon, T., Arvesen, A. & Hertwich, E. G. (2017). Life cycle assessment demonstrates environmental co-benefits and trade-offs of low-carbon electricity supply options. *Renewable and Sustainable Energy Reviews, 76*(C), 1283–1290.

International Energy Agency. (2014). Statistics. https://www.iea.org/statistics/balances/

International Energy Agency. (2016). *World energy outlook 2016.* https://www.iea.org/newsroom/news/2016/november/world-energy-outlook-2016.html

International Energy Agency. (2017). *What is energy security?* https://www.iea.org/topics/energysecurity/whatisenergysecurity/

Kaundinya, D. P., Balachandra, P., & Ravindranath, N. H. (2009). Grid-connected versus stand-alone energy systems for decentralized power—a review of literature. *Renewable and Sustainable Energy Reviews, 13*(8), 2041–2050.

Lampert, D. J., Lee, U., Cai, H., & Elgowainy, A. (2015). Analysis of water consumption associated with hydroelectric power generation in the United States. Energy Systems Division, Argonne National Laboratory. https://greet.es.anl.gov/publication-water-hydro

Macknick, J., Newmark, R., Heath, G., & Hallett, K. C. (2012). Operational water consumption and withdrawal factors for electricity generating technologies: a review of existing literature. *Environmental Research Letters, 7*(4), 45802.

McCall, J., Macknick, J., & Hillman, D. (2016). *Water-related power plant curtailments: An overview of incidents and contributing factors.* National Laboratory Report NREL/TP-6A20-67084. Golden, CO: US Department of Energy. https://www.nrel.gov/docs/fy17osti/67084.pdf

Meldrum, J., Nettles-Anderson, S., Heath, G., & Macknick, J. (2013). Life cycle water use for electricity generation: A review and harmonization of literature estimates. *Environmental Research Letters, 8*(1), 15031.

Moomaw, W., Yamba, F., Kamimoto, M., Maurice, L., Nyboer, J., et al. (2011). Introduction. In *IPCC special report on renewable energy sources and climate change mitigation* (pp. 161–208). Cambridge: Cambridge University Press.

Sathaye, J., Lucon, O., Rahman, A., Christensen, J., Denton, R., et al. (2011). Renewable energy in the context of sustainable development. Physics Faculty Publications, Paper 1. University of Dayton eCommons. https://ecommons.udayton.edu/phy_fac_pub/1/

Solley, W., Pierce, R., & Perlman, H. (1998). *Estimated use of water in the United States in 1995.* US Geological Survey Circular 1200.

Sullivan, J. L., Clark, C., Han, J., Harto, C., & Wang, M. (2013). Cumulative energy, emissions, and water consumption for geothermal electric power production. *Journal of Renewable and Sustainable Energy*, *5*(2), 23127.

Sullivan, J. L., & Wang, M. Q. (2013). Life cycle greenhouse gas emissions from geothermal electricity production. *Journal of Renewable and Sustainable Energy*, *5*(6), 63122.

Tidwell, V. C., Macknick, J., Zemlick, K., Sanchez, J., & Woldeyesus, T. (2014). Transitioning to zero freshwater withdrawal in the U.S. for thermoelectric generation. *Applied Energy*, *131*, 508–516.

Turconi, R., Boldrin, A., & Astrup, T. (2013). Life cycle assessment (LCA) of electricity generation technologies: Overview, comparability and limitations. *Renewable and Sustainable Energy Reviews*, *28*, 555–565.

US Energy Information Administration. (2017a). How much electricity is lost in transmission and distribution in the United States? https://www.eia.gov/tools/faqs/faq.php?id=105&t=3

US Energy Information Administration. (2017b). Levelized cost and levelized avoided cost of new generation resources in the annual energy outlook 2017. https://www.eia.gov/outlooks/archive/aeo17/pdf/electricity_generation.pdf

Van Vliet, M. T. H., Yearsley, J. R., Ludwig, F., Vögele, S. Lettenmaier, D. P., & Kabat, P. (2012). Vulnerability of US and European electricity supply to climate change. *Nature Climate Change*, *2*(9), 676–681.

CHAPTER 7

Domestic Water Use

7.1 Introduction

In 2010, the United Nations declared that access to safe drinking water and proper sanitation was a basic human right. In 2017, the UN estimated that 3 out of 10 people around the globe lacked access to safe and readily available drinking water at home and almost 6 in 10 lacked access to safely managed sanitation. Access to safe drinking water varies internationally between developed and developing countries and within nations between urban and rural residents. One of the **Sustainable Development Goals (SDG 6.1)** adopted by the UN in 2015 is to have safe and affordable drinking water for all by 2030. For a water utility to be considered safely managed, water free of fecal and priority contaminants must be available on the premises when needed (World Health Organization, 2017). Priority contaminants include chemical elements such as lead and arsenic as well as manufactured chemicals such as tetrachloroethylene, which is used widely for dry cleaning.

Safe and affordable drinking water falls under the domestic water use category. Domestic water use includes indoor water for drinking, cooking, bathing, flushing toilets, and so forth, and outdoor water for lawns and gardens. Domestic water can be delivered to the premises by a municipal water supply (also known as a public water supply) or can be self-supplied (e.g., by water from an individual well). It can be potable or nonpotable. *Potable water* is water that is considered to be safe for human consumption. *Nonpotable water*, such as water from untreated river water or rainwater, is not considered to be fit for human consumption. Nevertheless, nonpotable water can be an important water resource for uses of water not associated with human consumption, such as water for flushing toilets and water for outdoor landscaping. Domestic water can come from a variety of sources—surface

Box 7.1 Components of Municipal Water Supply and Treatment

Water source. A municipal supply can be from either surface water or groundwater or a combination of the two. The source water is referred to as raw water. An adequate source will supply enough water for all purposes, including domestic use, industrial use, and fire control use. An adequate source will provide sufficient water to cope with periods of drought.

Treatment of raw water. The source water is delivered by pipes to a water treatment plant. Two essential steps of treatment are filtration and disinfection. Passing raw water through a sand filter, as an example, will remove particles that have been suspended in the water. Disinfection is a step to kill harmful microorganisms and thereby render the water safe for drinking.

Water distribution system. After treatment, the finished water moves to a system of storage tanks and pipes and ultimately is delivered to households and businesses in the municipality.

Sewerage. After the finished water has been used for cleaning, cooking, bathing, flushing toilets, and so forth, it moves through sewer pipes to a wastewater treatment facility.

Wastewater treatment. Wastewater treatment typically involves several steps. Screens are used to remove larger solid material including dirt and sand that may have washed off city streets. Sedimentation tanks are used to coagulate and settle out fine solid material into sludge at the bottom of the tank. Chemical or biological treatment is then used to remove undesirable chemicals from the water before it is discharged from the treatment plant back to nature.

water, **groundwater**, desalinated sea water, and water previously used but treated so it can be recycled.

Municipal water supplies will play a large part in meeting **SDG 6.1**. A little more than half of the global population resides in urban areas, and trends indicate that urban populations will increase to two-thirds of the total population by 2050. There are a number of components required to ensure a safely managed **municipal water system** (Box 7.1). First, of course, there must be an adequate supply of fresh water at the location of the municipality. This requirement implies availability through time; that is, water must be available during all periods of a year and even during extended droughts. Second, water treatment must be implemented to meet the basic standards of adequate drinking water quality. And finally, there must be a distribution system to deliver treated water to households and a system to remove wastewater and treat it so it can be returned safely to the environment (Figure 7.1).

7.2 Reliability

A water supply is considered to be reliable when water generally is available to meet demands fully. Water supply systems are not perfectly reliable in that they may fail to deliver water under

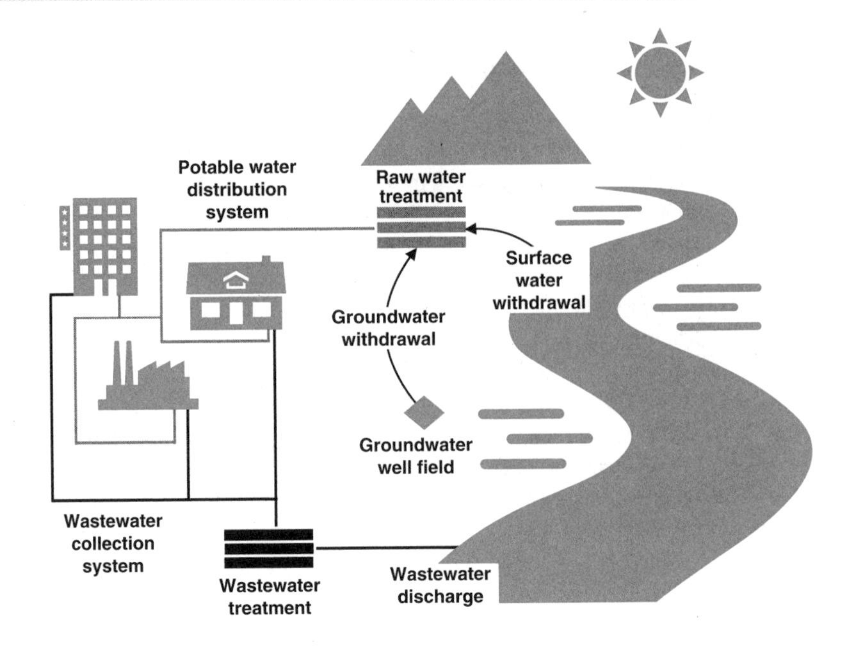

Figure 7.1 Schematic diagram of a municipal water withdrawal, treatment, and distribution system.

extreme conditions. Water supply systems can be centralized or decentralized. Like our energy systems (Chapter 6), centralized water systems acquire, treat, and distribute water through a network to end users.

Decentralized water systems are stand-alone water systems, often connecting a small number of water users to a single water source. A good example of a decentralized water supply system is a **groundwater** well responsible for providing a household with water. Although decentralized systems may reduce conveyance losses, centralized systems can use a number of water sources—such as **groundwater**, surface water, desalinated water—and **infrastructure** to store large quantities of water (such as in reservoirs) to increase resilience during times of unusual, long-term drought. Municipalities typically develop centralized systems to be as reliable as possible.

At a very basic level, the adequacy of a water supply is constrained by hydrology. How much water is available to cities? One way to estimate water availability is to compute a water budget (Chapter 1) for each location on the globe and include computations to route water not used locally to downstream locations in the catchment or drainage basin area (Chapter 2). Sustainable **groundwater** use also can be considered by estimating water pumped, recharge rates to **aquifers**, and discharge rates to connected surface waters

Box 7.2 New York City Water Supply

New York City does not have ready access to a large supply of fresh water within its borders. The need to reach outward and transfer water from elsewhere into the city was recognized quite early in the city's history. In 1842, water was delivered to the city from a reservoir constructed on the Croton River, 40 miles from the city through a covered aqueduct. The Croton system continued to be expanded but even that proved inadequate to supply the growing New York metropolis.

New York City next acquired rights to water in the Catskill Mountains, and in 1927 water was delivered to the city from two reservoirs constructed in the Catskills through a 92-mile aqueduct. Finally, in 1964 water from the Delaware River began to be delivered to New York City though a 105-mile aqueduct.

These three inter-basin transfer systems—the Croton, Catskill, and Delaware—constitute the current water supply for the city. The bulk of the water supply (90% or more) is from the combined Delaware–Catskill sources, which are of such good quality that they are delivered to consumers without filtration. The success of the New York City development of its water supply by transferring water from other basins has proved to be a model for many other cities around the world.

(Chapter 3). But as cities grow and collect resources from the hinterlands, the reach for water often extends beyond the drainage basin in which they are located (Chowdhury et al., 2013). That is, cities depend on *inter-basin transfers of water*, so a full accounting of water available to a city must include these transfers (Box 7.2).

How can adequacy of a water supply for a city be judged? One metric that often is used for water stress for surface water sources is when water use exceeds 40% of water available. An indicator of **groundwater** stress is when withdrawals exceed recharge rates adjusted for required environmental discharges to rivers—essentially an estimate of the **safe yield** of an **aquifer** (Chapter 3). One-quarter of large urban centers in the world can be considered water stressed relative to surface water and **groundwater** availability measures (McDonald et al., 2014).

A water stress analysis indicates whether the current water delivery is in a sustainable range. It does not reflect whether current municipal deliveries are adequate. For example, consider the case of two cities, Los Angeles in the United States and Nairobi in Kenya. Los Angeles, at 34 degrees N latitude and at an elevation of 82 meters, has a mild climate with an average annual temperature of 17.7°C and an average annual rainfall of 475 millimeters, essentially all of which occurs during the Northern Hemisphere winter. Nairobi, at 1.3 degrees S latitude and at an elevation of 1,690 m, has a subtropical climate with an average annual temperature of 19°C and an average annual rainfall of 869 mm, which occurs mainly in two seasons (March–May and November–December). Currently, Los Angeles is considered water stressed, and Nairobi is not considered to be water stressed (McDonald et al., 2014). Yet the population of Los Angeles currently has continuous access to safe drinking water while many residents of Nairobi currently suffer intermittent access and endure frequent issues with poor water quality.

7.3 Economics

Here we focus on the supply aspects, and on the needs for collection, storage, and distribution of fresh water (see Figure 7.1) because water supply **infrastructure** requires significant financial resources. In terms of water supply, a complete urban water management system would collect and store enough water to meet demands (requiring millions to a few billions of US dollars), treat the water to achieve a high degree of purity (millions of US dollars), distribute the water to various users (millions of US dollars), and then collect the water after use and treat the used water for reuse by downstream users (millions of US dollars). In addition to the supply aspects, a system would also have to deal effectively with material that is removed in wastewater treatment and with stormwater runoff (millions of US dollars).

The approach to urban water supply relies heavily on investment in **infrastructure**. Reservoirs are constructed to collect and store water, **groundwater** pumping fields are established, central water treatment plants are operated, and pipe networks are constructed to bring water from the source to the treatment plant and to deliver piped water to customers. In many instances, and particularly in arid areas, the source water can be long distances from the urban center, and the conveyance losses can be substantial (Box 7.3).

The development of **infrastructure** implies large capital investment, so predictably the success in providing a population with access to adequate, high-quality water is much greater in high-income countries than in developing countries. For example, the Los Angeles Department of Water and Power supplies 3.9 million people at an average rate of more than 400 liters per capita per day (Lpcd) (Los Angeles Department of Water and Power, n.d.). The Nairobi City Water and Sewerage Company in Kenya supplies ap-

Box 7.3 Municipal Water in Tucson, Arizona

The city of Tucson, located in the southern part of Arizona, uses water from a variety of sources: the Colorado River (a distant source) through the Central Arizona Project (CAP), local groundwater, and treated wastewater (reuse). A snapshot of the water portfolio in 2005 shows that CAP water and groundwater are approximately equal primary sources (Figure B7.3.1). The overall conveyance losses from these primary sources amount to more than 20%. There is also a significant energy requirement for water transported over long distances (Perrone, Murphy, & Hornberger, 2011).

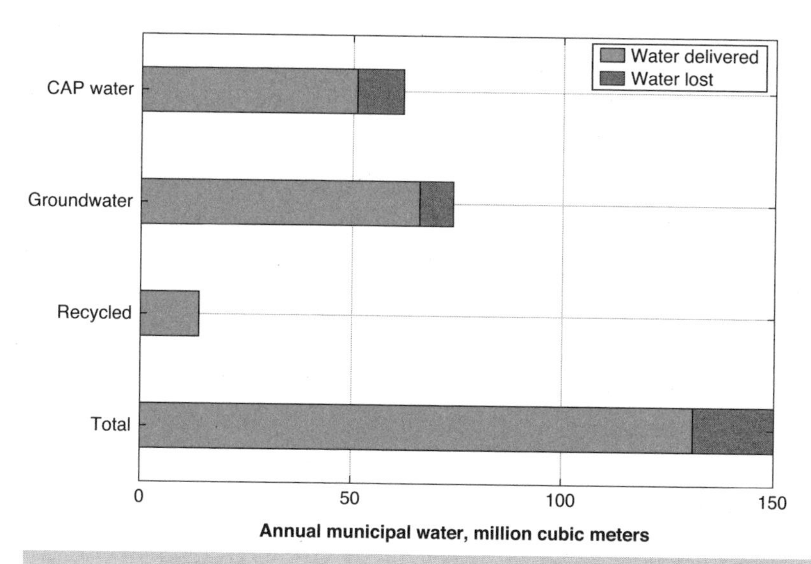

Figure B7.3.1 Municipal water portfolio for Tucson, Arizona, in the United States. The blue bars represent delivered water, and the gray bars represent losses in the system. CAP refers to the Central Arizona Project. Data from Perrone et al. 2011.

proximately 500,000 m³/day to 3.8 million people (Mafuta et al., 2011; Nairobi City Water and Sewerage, 2017) for an average rate of about 140 Lpcd; this figure does not take water losses into account, which may approach 50%. Only about half of the populace of Nairobi is served by piped water whereas essentially 100% of the populace of Los Angeles has piped water.

The need for new **infrastructure** investments is clear, especially given the pace of urbanization in the world. There are major needs to upgrade aging pipes and other parts of the water supply **infrastructure** in both developed and developing countries. Losses from leaking pipes in the United States range from about 21 to 134 liters per day per service connection (Ress & Roberson, 2016) for a total of about 17% of total municipal water delivered by municipal services across the country (Baird, 2011). The corresponding figures are much higher for low- and medium-income countries. For example, water lost through leaking pipes and illegal connections in Nairobi may be 50% of drinking water delivered to the city (Engel et al., 2011).

The response of cities to the challenge of meeting municipal water demands in a sustainable way will vary with circumstances. The approach used in the past was to make massive investments in dams, pipelines, and **groundwater** well fields and to seek water supplies from ever more distant places in the hinterland surrounding cities. The rate at which the world is urbanizing, global population is expanding, and climate change is occurring gives a clear signal that this approach will not be sufficient in the future for many cities. The supply of water will certainly have to be increased for growing cities, but the efficiency of water used will have to be increased, and the demand for water will have to be managed as well.

In high-income countries it is likely that a relatively broad portfolio of strategies will be needed. These strategies could include construction of additional pipelines, reservoirs, well fields, and desalination plants, as well as the adoption of new technologies to improve efficiency, limit water loss, and reuse water. Behavioral strategies that focus on water conservation will likely play a fundamental role, too. For example, Los Angeles anticipates a modest population growth of 0.45% per year giving a projected population in 2035 of 4.35 million. The city anticipates essentially no increase or a very small increase in demand for water by 2035 (Los Angeles Department of Water and Power, 2016). Nevertheless, there are major changes planned in how the necessary water is obtained. Historically, Los Angeles has relied on water that is transported for hundreds of kilometers: the Los Angeles Aqueduct transfers water from the east side of the Sierra Nevada Mountains; the State Water Project transfers water from northern California to southern California; and the Colorado River Aqueduct transfers water from Arizona. The plan for the future is to capture local stormwater runoff for use (including to recharge aquifers), increase the use of recycled water, and institute additional conservation measures. The goal is to greatly reduce the dependence on long-distance water transfers from about 87% to 53% (Los Angeles Department of Water and Power, 2016). The example illustrates that even though cities in high-income countries face serious issues with respect to satisfying municipal water demands, the economic resources, the institutions, and the governance structures are in place to respond to challenges.

The situation in low-income countries is, in general, much different. First, there is a current shortfall in water supplied that requires significant investment to overcome. Second, urban centers in these countries are growing rapidly, and simply keeping pace with growth is problematic. Consider once again, the city of Nairobi in Kenya. The current population is about 3.8 million, and the projected population for 2035 ranges from 6.4 to 7.8 million (World Bank, 2011). Even the higher estimate of 7.8 million people is for a growth rate of 3.5% per year. The recent growth rate has been 4% per year, so the population in 2035 could easily be more than 8 million. The corresponding estimates of water supply for the World Bank population estimates are from 1.2 to 1.4 million m^3/day, more than double the current supply of about 550,000 m^3/day. In 2015, the so-called Northern Collector Project, a US $1 billion construction to transfer water from rivers in the Kenyan highlands to Nairobi, was started. The water plan appears to be focused almost exclusively on increasing the supply of water, not on the delivery system to consumers (Blomkvist & Nilsson, 2017).

New **infrastructure** investments to secure increased supplies of water are a critical need in Nairobi as well as in many other cities. In addition, there is a critical need to manage demand (Ledant, 2013) so that less water per capita is needed. Conservation is especially important given that water supplies for the projected population growth for Nairobi beyond 2035 (Hoornweg & Pope, 2017) simply cannot be achieved through ever more massive projects to secure new sources.

7.4 Environment

Global population is projected to increase significantly this century (Chapter 9). The number of people living in urban areas is expected to increase by about 1 billion by 2030 (United Nations, 2016) with much of the increases in Asia and Africa. In 2016, there were 45 cities with a population between 5 and 10 million; 29 more cities will see their populations grow to 5 to 10 million by 2030. These statistics do not imply that we will "run out of water." Water is a renewable resource and can be provided to cities so long as the planning for use and management of the resource is sound. The challenge in the face of competing demands on water resources will require the use of advances in technology and adjustments in how institutions handle the new water systems (Larsen et al., 2016). A particular issue will be to ensure that water needs for ecosystems will be met alongside the water needs of humans (Chapter 8).

Domestic use of water can impact ecosystems in several ways. Perhaps most obvious is the potential for degradation of water quality by contaminants in the effluents discharged once domestic water has been used. Wastewater treatment plants can ensure that degradation does not occur, but while such plants are the norm in developed countries, they are not ubiquitous in the developing world. For example, Los Angeles has advanced wastewater treatment plants that deliver very high-quality water suitable for reuse (Los Angeles Department of Water and Power, 2016), whereas Nairobi struggles to reach minimal levels of treatment (Mafuta et al., 2011; Nairobi City Water and Sewerage, 2017). The construction of water supply **infrastructure** also has an impact on ecosystems. As one example, the construction of reservoirs inundates lands and displaces terrestrial ecosystems, restricts the migration of anadromous fish (e.g., salmon) upstream during spawning seasons, and changes the temperature of water released downstream, affecting fisheries.

7.5 Water Use

The determination of what constitutes an "adequate" water supply depends on context. To live, a person needs about 2 liters per day of drinking water. But that is unrealistic as an adequacy measure in all but an extreme, short-lived case. People need safe water for food preparation, for bathing, and for laundering at a base minimum, which brings the minimum to tens of Lpcd—perhaps 50 Lpcd (Gleick, 1996). Estimates for minimal adequate municipal deliveries (to businesses, institutions, and households) must include water needed to have a viable socioeconomic structure. A value of 135 Lpcd has been suggested as a reasonable minimum (Chenoweth, 2008).

The actual amount of water that people use varies greatly. Municipal water withdrawals range from less than 20 Lpcd to more than 500 Lpcd (Figure 7.2). Among 163 countries for which withdrawal estimates are available (FAO, 2016), 20 withdraw more

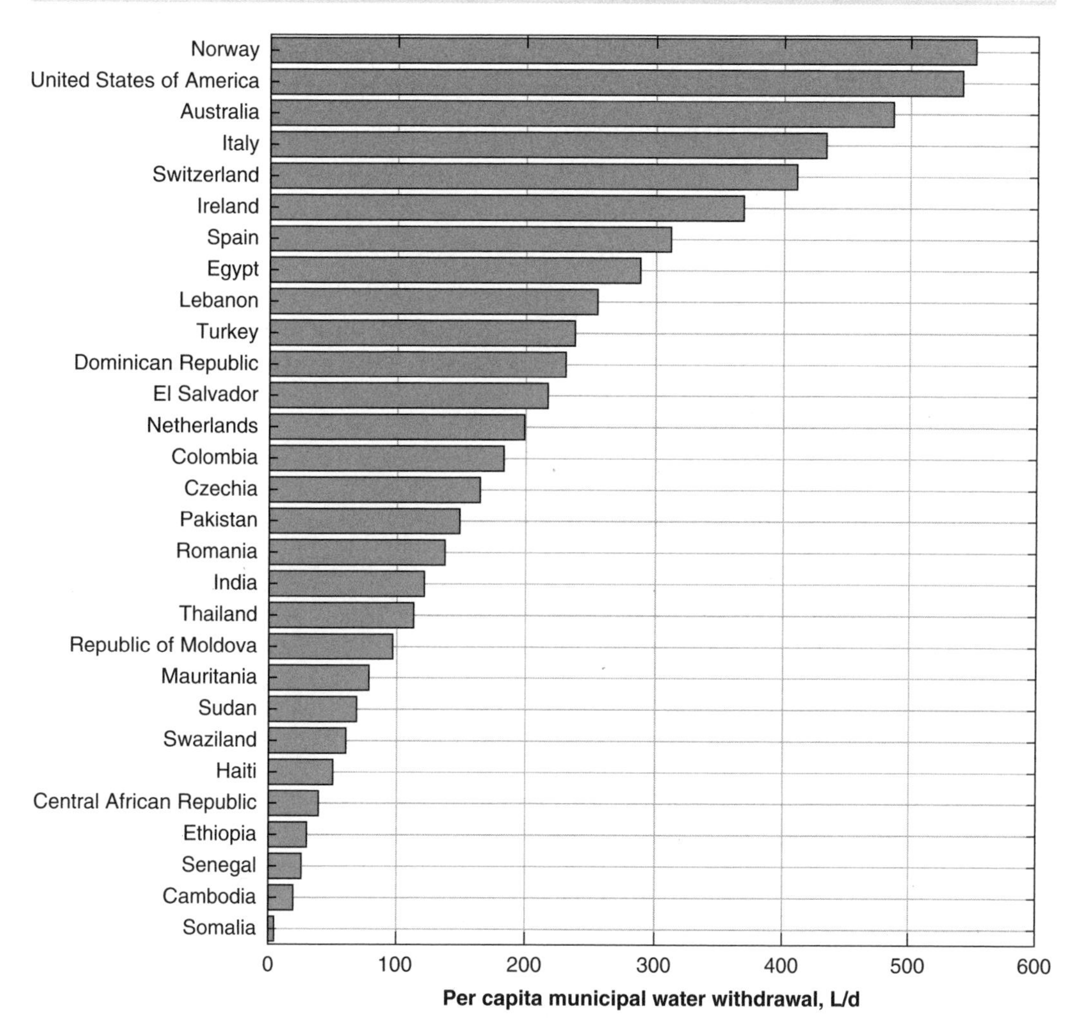

Figure 7.2 Per capita municipal water withdrawals for selected countries that span the range of reported values. Data from FAO 2016.

than 400 Lpcd and 25 withdraw less than 20 Lpcd. The UN's **Sustainable Development Goals** for water are noble, but the difficulty of the path for achieving success with respect to municipal water supplies is clear (Satterthwaite, 2016), especially in the face of increasing urbanization worldwide.

7.6 Concluding Remarks

Urban water management has historically relied on the use of water sources from wherever feasible and the development of large-scale **infrastructure** to collect, store, treat, and deliver the water. That development logic persists to a large extent throughout the world, even in relatively water-poor areas of the United States and despite the fact that adopting water conservation measures would be both economically and environmentally beneficial (Hess et al., 2016). Looking to the future, there will be a need to develop and implement a variety of new technologies. It will be equally critical to manage the demand for water through changes by individuals, institutions, and organizations, to adopt appropriate governance structures, and to find ways to overcome inertia to change in urban water management institutions (Daigger, 2011; Larsen et al., 2016; Perrone & Hornberger, 2014).

7.7 Key Points

- One of the UN's **Sustainable Development Goals** is for all people in the world to have access to an adequate supply of safe drinking water by 2030. (Section 7.1)
- The availability of water to a city depends on both the adequacy of the source of water and the **infrastructure** and the institutional structure to deliver it. (Section 7.2)
- There is a need for major investments in **infrastructure** for cities, both to renew aging **infrastructure** and to develop new systems. (Section 7.2)
- To achieve the goal of having access to safe water for an increasing urban world, demand for water will have to be managed and good governance will have to be instituted. (Section 7.3)
- An estimate for minimum water needed for cities with a good level of economic activity is about 135 liters per capita per day (Lpcd). (Section 7.5)
- Average water use per person varies widely, from several hundred Lpcd in high-income countries to less than 20 Lpcd in some developing countries. (Section 7.5)

7.8 Example Problems

Problem 7.1. According to the World Wildlife Fund (Engel et al., 2011), the city of Karachi, Pakistan, has a population of 18 million. Municipal water is delivered at an average rate of 33 m^3s^{-1}.

a. About 35% of the water is lost to leakage or unauthorized connections. What is the average rate of municipal water delivery in liters per capita per day (Lpcd)? (There are 1,000 liters per cubic meter.)

b. How does this compare to the suggested minimum requirements?

Problem 7.2. The average rainfall in Karachi is 167.6 mm/year and the area of the city is 3,527 km^2.

a. What is the volume of rain that falls on the city expressed as L/day?

b. If 25% of the rain could be harvested and stored safely, how many Lpcd could be supplied?

Problem 7.3. A new project underway by the Karachi Water and Sewerage Board, the K-4 Project, is designed to bring an additional 1,200 m^3/day to the city.

a. Assume a goal of 135 Lpcd and calculate how many people could be served by this additional water. (Assume 35% of the water is lost to leakage or unauthorized connections.)

b. If Karachi grows to a population of 32 million in 2050 (Hoornweg & Pope, 2017), what could the current supply (Problem 7.1) and the additional supply from K-4 be expected to yield in Lpcd if the losses remain at about 35%?

c. How would the expected yield change if losses are reduced to 15% and the population is 32 million?

d. How would the expected yield change if the losses are reduced to 15% and the population grows to 49 million by the year 2100?

7.9 Suggested Reading

Larsen, T. A., Hoffmann, S., Lüthi, C., Truffer, B., Maurer, M., et al. (2016). Emerging solutions to the water challenges of an urbanizing world. *Science, 352*(6288), 928–933.

7.10 References

Baird, G. M. (2011). Who stole my water? The case for water loss control and annual water audits. *Journal American Water Works Association, 103*(10), 22–30.

Blomkvist, P., & Nilsson, D. (2017). On the need for system alignment in large water infrastructure: Understanding infrastructure dynamics in Nairobi, Kenya. *Water Alternatives, 10*(2), 283–302.

Chenoweth, J. (2008). Minimum water requirement for social and economic development. *Desalination, 229*, 245–256.

Chowdhury, F., Lant, C., & Dziegielewski, B. (2013). A century of water supply expansion for ten U.S. cities. *Applied Geography, 45*, 58–76.

Daigger, G. (2011). Sustainable urban water and resource management. *The Bridge, 41*, 13–18.

Engel, K., Jokiel, D., Kraljevic, A., Geiger, M., & Smith, K. (2011). *Big cities, big water, big challenges: Water in an urbanizing world.* Berlin:

WWF. https://www.wwf.se/source.php/1390895/Big%20Cities_Big%20Water_Big%20Challenges_2011.pdf

Food and Agriculture Organization (FAO). (2016). AQUASTAT - Main Database. http://www.fao.org/nr/water/aquastat/data/query/index.html

Gleick, P. H. (1996). Basic water requirements for human activities: Meeting basic needs. *Water International, 21*(2), 83–92.

Hess, D. J., Wold, C. A., Hunter, E., Nay, J., Worland, S., et al. (2016). Drought, risk, and institutional politics in the American Southwest. *Sociological Forum, 31*, 807–827.

Hoornweg, D., & Pope, K. (2017). Population predictions for the world's largest cities in the 21st century. *Environment and Urbanization, 29*(1), 195–216.

Larsen, T. A., Hoffmann, S., Lüthi, C., Truffer, B, Maurer, M., et al. (2016). Emerging solutions to the water challenges of an urbanizing world. *Science, 352*(6288), 928–933.

Ledant, M. (2013). Water in Nairobi: Unveiling inequalities and its causes. *Les Cahiers d'outre-Mer, 66*(263), 335–348.

Los Angeles Department of Water and Power. (2016). *Urban Water Management Plan 2015.* https://planning.lacity.org/eir/CrossroadsHwd/deir/files/references/M217.pdf

Los Angeles Department of Water and Power. (n.d.). Facts and figures. https://www.ladwp.com/ladwp/faces/ladwp/aboutus/a-water/a-w-factandfigures?_adf.ctrl-state=11ai1eeedy_17&_afrLoop=598675716248388

Mafuta, C., Formo, R. K., & Nellemann, C. (2011). *Green hills, blue cities: An Ecosystems Approach to water resources management for African Cities.* United Nations Environment Programme/GRID-Arendal. https://www.unenvironment.org/resources/report/green-hills-blue-cities-ecosystems-approach-water-resources-management-african

McDonald, R. I., Weber, K., Padowski, J., Flörke, M., Schneider, C., et al. (2014). Water on an urban planet: Urbanization and the reach of urban water infrastructure. *Global Environmental Change, 27*(1), 96–105.

Nairobi City Water and Sewerage. (2017). Nairobi City Water and Sewerage Company (NCWSC). https://www.nairobiwater.co.ke/index.php/en/

Perrone, D., & Hornberger, G. M. (2014). Water, food, and energy security: Scrambling for resources or solutions? *WIREs Water, 1*(1), 49–68.

Perrone, D., Murphy, J., & Hornberger, G. M. (2011). Gaining perspective on the water-energy nexus at the community scale. *Environmental Science and Technology, 45*(10), 4228–4234.

Ress, E., & Roberson, J. A. (2016). The financial and policy implications of water loss. *Journal American Water Works Association, 108*(2), E77–E86.

Satterthwaite, D. (2016). Missing the Millennium Development Goal targets for water and sanitation in urban areas. *Environment and Urbanization, 28*(1), 99–118.

United Nations. (2016). *The world's cities in 2016—data booklet.* ST/ESA/SER.A/392. Department of Economic and Social Affairs, Population Division. http://www.un.org/en/development/desa/population/publications/pdf/urbanization/the_worlds_cities_in_2016_data_booklet.pdf

World Bank. (2011). *Feasibility study and master plan for developing new water sources for Nairobi and satellite towns.* Republic of Kenya, Ministry of Water and Irrigation. http://documents.worldbank.org/curated/en/169681468272734316/Feasibility-study-and-master-plan-for-developing-new-water-sources-for-Nairobi-and-satellite-towns

World Health Organization. (2017). *Safely managed drinking water—thematic report on drinking water 2017.* Geneva, Switzerland. https://data.unicef.org/wp-content/uploads/2017/03/safely-managed-drinking-water-JMP-2017-1.pdf

CHAPTER 8

Environmental Water Use

8.1 Introduction

The connections between water and life are ubiquitous and without substitutes. Life on Earth evolved once liquid water formed, and Earth's evolution has been shaped by the flow of water ever since. Water is essential, not only for direct human well-being (Chapter 7) but for all living organisms. Throughout their linked life cycles, aquatic and terrestrial flora and fauna engage in a suite of processes that benefit humans in many ways. These processes are referred to as **ecosystem services.**

Ecosystem services are critical to maintaining a stable and robust environment hospitable to humans. **Ecosystem services** are of several different types (Table 8.1). Ecosystems provide things necessary for humans and thus deliver provisioning services. Provisioning services are essential for food production because the bulk of staple products are from managed agroecosystems and aquatic environments. Ecosystems also help to maintain stable environmental characteristics, and thus convey regulating services. A primary regulating service in aquatic systems is the removal of contaminants from water—wetlands often are referred to as the kidneys of Earth. Ecosystems support a "home" for the complex web of species that live therein, providing habitat services.

The habitat services provided by freshwater systems are astounding. Although lakes, rivers, and wetlands make up a small fraction of total water (see Table 1.1), and a correspondingly small fraction of the surface area of Earth, they are home to more than 9% of the total number of animals known to biologists (Turak et al., 2017). Finally, ecosystems deliver a host of cultural services such as environments that attract tourists and that inspire artists. The category of cultural services encompasses a wide variety of benefits

Table 8.1 Aquatic ecosystem services.

Type of service	Examples
Provisioning	Drinking water supply
	Nondrinking water (agriculture, industry, energy production, etc.)
	Fisheries (food provisioning)
Regulating	Water purification
	Flood protection
	Erosion and sedimentation control
Habitat	Maintaining biodiversity
	Nurseries for fish
	Protection of endangered species
Cultural	Recreation
	Aesthetic appreciation
	Spiritual associations

that are difficult or impossible to measure quantitatively. Included in the category are things like recreation, aesthetic enjoyment, and spiritual experiences.

In 2000, the United Nations initiated the Millennium Ecosystem Assessment and engaged experts from around the world in the evaluation process. A general conclusion was that humans have seriously compromised the Earth's **ecosystem services** to the point that there may be very severe repercussions in the future. In particular, the group that reported on freshwater ecosystems concluded that "the condition of inland waters . . . has been compromised by the conventional sectoral approach to water management, which, if continued, will constrain progress to enhance human well-being" (Millennium Ecosystem Assessment, 2005).

8.2 Reliability

Stresses on freshwater ecosystems can be evaluated in terms of water quantity (too much water or too little water), water quality, and changes in the physical habitat. Natural stresses on freshwater ecosystems arise from floods and droughts, which can affect these ecosystems' quantity, quality, and physical habitat. For example, a severe drought greatly reduces water flow in streams, alters the temperature and chemical composition of the water, and restricts the available habitat for fish and other aquatic organisms. Although impacts on ecosystems can be severe (e.g., a major fish die-off due to drought), our expectation is that, in freshwater ecosystems that are not highly impacted by humans, the system will be resilient and return to a stable state in time.

When we are thinking about freshwater ecosystems, we often concentrate on surface water resources (Chapter 2). Wetlands and riparian zones, or the areas that interface surface waters and land, provide distinct ecosystems for plants and animals. From earlier chapters, we know that the interconnection between surface water and **groundwater** resources can be significant, such that **groundwater** withdrawals affect water flow to streams and wetlands and can even drain them completely under severe circumstances (e.g., see Box 3.2). **Groundwater** resources also provide direct benefit to many plants and animals. Ecosystems that require **groundwater** flows for some or all of their survival requirements are referred to as **groundwater-dependent ecosystems (GDEs)**. The need to manage **groundwater** withdrawals to protect **GDEs** has been recognized around the world, and various regulations have been put in place to guide planning and management efforts (Rohde et al., 2017).

The appropriation of water by humans for various purposes can lead to chronic stresses on freshwater ecosystems. Various water uses, including environmental flows to sustain **ecosystem services**, are interdependent. One convenient way to keep perspective on these interrelationships is to consider the framework known as the water-energy-food nexus (Perrone & Hornberger, 2014). Irrigated agriculture represents the world's largest consumptive use of water—including for food crops, for biological materials, and for biofuels—and it affects water quantity in freshwater systems directly (Figure 8.1A). The widespread application of agricultural chemicals including fertilizers, herbicides, and pesticides in agricultural systems impacts water quality as well. Water also is used for energy production and industry (see Figure 8.1B). Although the use of water for cooling in thermoelectric plants and for industrial use such as cleaning is largely nonconsumptive, the water returned to surface water is at an elevated temperature or contains contaminants and can be a stressor for freshwater ecosystems (see Figure 8.1B). The construction of dams for hydroelectric power production creates new reservoir habitat for aquatic organisms, but it changes the natural flow pattern in downstream rivers on time scales ranging from daily to seasonal to interannual. Urbanization changes land use substantially, with efficient drainage **infrastructure** that leads to large changes in the timing of runoff to streams and rivers (see Figure 8.1C). The municipal use of water

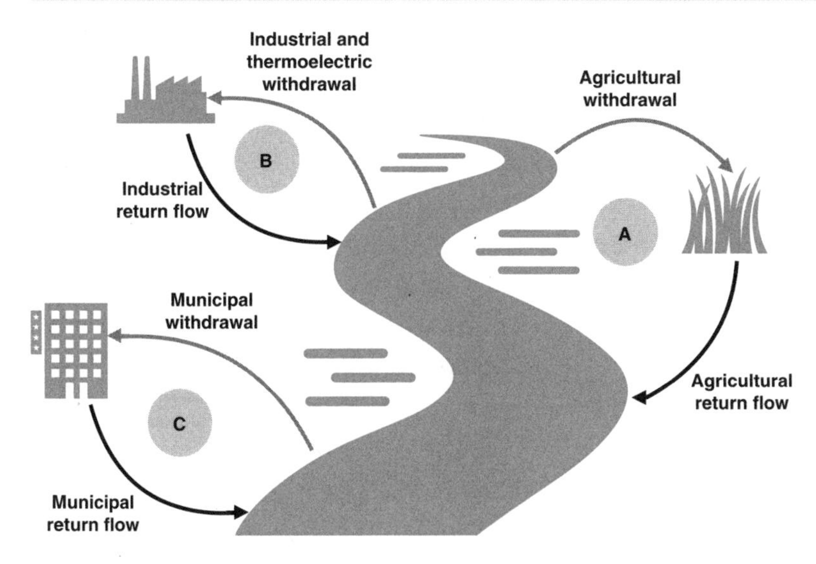

Figure 8.1 Environmental water flows are affected by (A) agricultural, (B) thermoelectric and industrial, and (C) urban water uses and land changes. Water quantity and quality are affected.

is tiny compared to agricultural use, but effluent from waste streams has a major impact on water quality (see Figure 8.1C). Land clearing for agriculture, for industry, and for urban development leads to erosion, which elevates the sediment concentration in surface waters and also can change the physical habitat when the sediment is deposited.

Freshwater ecosystems have been impacted by human appropriation of water resources for at least several centuries. Although there are well-documented disasters caused by human activities, globally the resilience of freshwater ecosystems has buffered impacts to a great extent. Over the past several decades, however, the stresses have increased and are expected to continue to increase and will potentially overwhelm freshwater ecosystems. The main drivers of change are population and economic growth, the concomitant increases in demand for agricultural and industrial water use (Chapter 9), and climate change (Chapter 10). Making good decisions about managing water resources will be critically important in coming decades.

8.3 Economics

Environmental water uses, such as water to support wetlands and maintain water quality, can be difficult to calculate, controversial to monetize or value, and not understood well by the public. Analyses of balancing water resource use to reverse long-term degradation require information on critical **ecosystem services**. Wise decisions require information about water use, the biophysical needs of ecosystems in terms of providing the desired services, the human actions that place stresses on ecosystems that affect services, the steps that can be taken to mitigate adverse impacts, and the value that various stakeholders place on the **ecosystem services** relative to alternative uses of the water. Approaches for evaluating **ecosystem services** relative to other uses of water are complex (National Research Council, 2005), and a comprehensive discussion is beyond our scope. We will focus on two examples to highlight how **ecosystem services** can affect decisions about the use of water resources.

A relatively straightforward example of placing an economic value on an **ecosystem service** is New York City (NYC), which derives its main water supply from the distant Catskill Mountains and the upper Delaware River Basin (see Box 7.2). The water quality is very high, so the water does not have to be filtered before it is delivered to customers. That is, the forested ecosystems in the Catskill–Delaware landscape supply the provisioning and regulating services of filtering the water to meet drinking water standards. In the 1980s, the US Environmental Protection Agency (EPA) was considering whether NYC should build a treatment plant to filter the water supply, a requirement of the Clean Water Act (Chapter 11). The cost would have been billions of dollars, with significant ongoing costs to operate the treatment facility. Instead, NYC negotiated an exception with the EPA in which they invested in maintaining and enhancing the **ecosystem services** provided by the watersheds that supply the water. These investments included purchasing land to ensure that the natural forests would not be converted to agriculture or other land uses that could potentially degrade the water quality and working with farmers in the region to adopt best practices to avoid causing water quality problems.

Another example is estimating the value of GDEs. Wetlands are one important category of **GDEs**. Given that over 50% of the world's wetlands have already been lost (Chapter 2) and that both urban and agricultural expansion continue to threaten destruction of wetlands, the value—both the strict economic value and the cultural value—of wetlands needs to be recognized. Consider wetlands in Beijing in the People's Republic of China (Zhang et al.,

2017). Wetlands in Beijing contribute provisioning services (inflow to reservoirs, recharge to **groundwater**), regulating services (flood protection, cooling, purification), and habitat services to a variety of plants and animals. An overall estimate of the total value of the wetlands is 360,000 Yuan per hectare per year, which is about US $60,000 per hectare per year. Such analyses indicate that there is direct economic value in **GDEs**, and this evidence may support management decisions that can protect against further degradation in the future.

8.4 Water Use

There are many attributes of water flows that affect **ecosystem services**. One major aspect is ensuring that low flows in rivers are maintained at levels that minimize damage to ecosystems. A certain level of water flow, often dubbed an environmental *low-flow requirement (LFR)*, is necessary to maintain the physical habitat. For example, if the flow drops below some threshold, water may become isolated in pools that are disconnected from the flowing water of a stream or river. This could have a negative impact on some aquatic species. In addition, low flows can lead to deterioration of water quality. River flows often are critically low in summer when high water temperatures are a constraint on ecosystem functioning, so water temperature is inversely correlated with the flow itself (Figure 8.2).

In recognition that "environmental flows are essential for freshwater ecosystem health and human well-being," the Brisbane Declaration (2007) set out an agenda for all countries to maintain and restore environmental flows. Objectives established by the United Nations in their **Sustainable Development Goals** explicitly support the Brisbane Declaration by incorporating environmental flows into water planning, stating the aim for 2020 is to "ensure the conservation, restoration and sustainable use of terrestrial and inland freshwater ecosystems and their services" (Goal 15.1). The need to attend to providing sufficient environmental flows is urgent. More than 50% of readily available fresh water is appropriated for human use, and about 65% of biodiversity is threatened (Vörösmarty et al., 2010). Nevertheless, achieving the proper balance between water appropriation for direct human use and water use for environmental flows is far from a simple task.

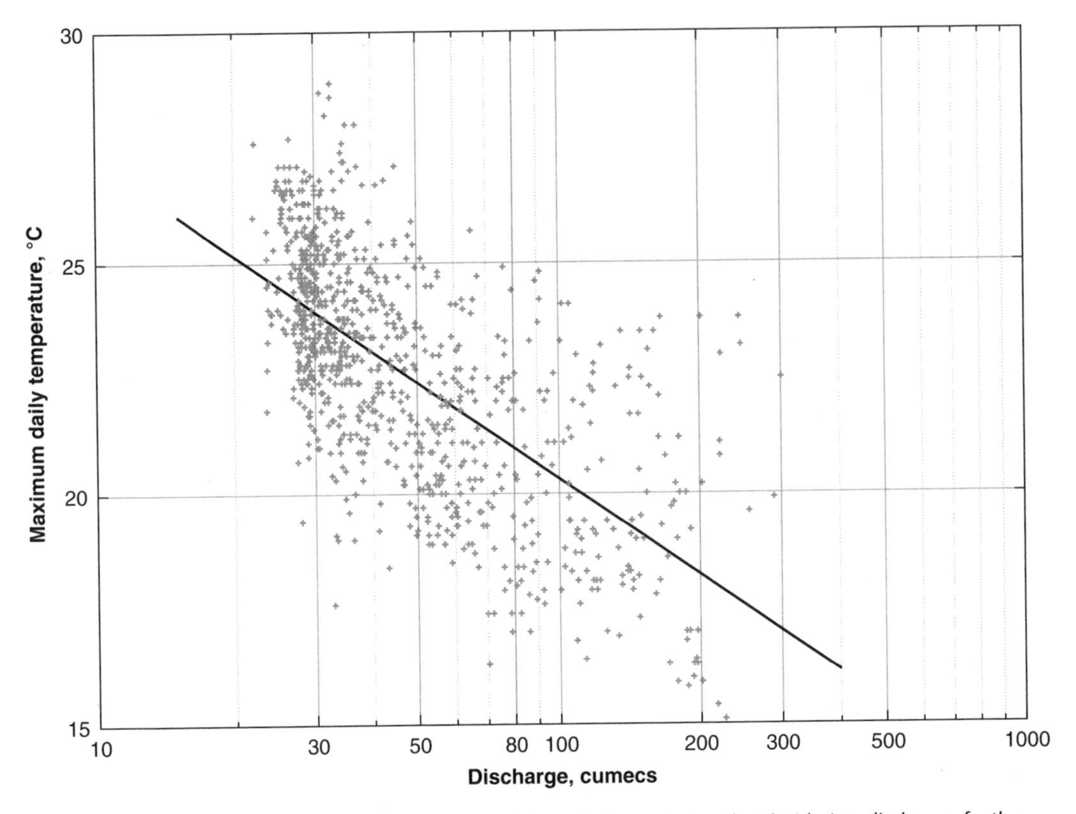

Figure 8.2 Water temperature for the months of July and August is inversely correlated with river discharges for the Chattahoochee River near Atlanta, Georgia. Data from US Geological Survey.

Trade-offs have to be made, and these trade-offs will depend on circumstances: the aim cannot be to restore pristine conditions except in rare cases. This indicates that different objectives will have to be set for each river or watershed or wetland. Because of the interrelationships among water quantity and quality and food and energy production (Perrone & Hornberger, 2014), efforts to plan and manage jointly across these sectors will be essential. Because water use trade-offs have typically been confrontational rather than collaborative, new institutions, including laws and regulations, will have to be devised. One extreme example can be to assign legal rights to rivers (O'Donnell & Talbot-Jones, 2018).

One of the methods that water resource managers use to determine LFRs is to consider the flow duration curve for rivers. Under natural or modestly modified conditions, the ecosystem in a river will have evolved and adapted to the prevailing flow regime. Acute stress may occur at low flows that occur only 5%

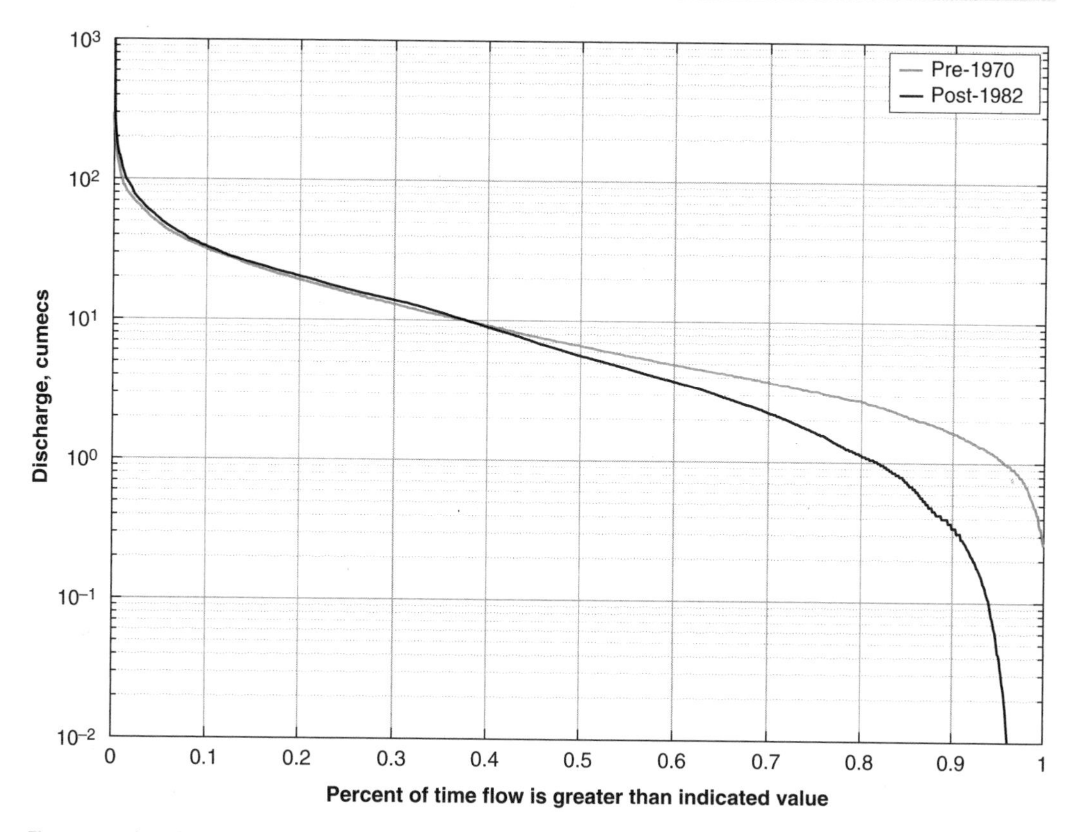

Figure 8.3 Flow duration curve for Spring Creek in southwest Georgia, United States. The pre-1970 flow duration curve was completely changed at the low-flow end of the curve after the large-scale development (post-1982 curve). Data from US Geological Survey.

or 10% of years. Flows at these levels would correspond to the discharge that is equaled or exceeded 95% or 90% of the time, respectively. Hydrologists represent these quantities as Q95 and Q90. For example, for the Harpeth River in Tennessee, Q90 is about 0.011 cumecs and Q95 is about 0.006 cumecs (see Figure 2.10).

Environmental LFRs must be considered in the context of the multiple uses of water that compete. Months of low flow in streams and rivers that are critical for ecosystems are also periods when water requirements for irrigation or for cooling thermoelectric plants also are present. If abstractions for, say, irrigation are not regulated, there can be significant impacts on low flows. For example, for Spring Creek, Georgia (see Box 3.2), the flow duration curve was completely changed at the low-flow end of the curve after the development of large-scale irrigation (Figure 8.3).

One approach that has been taken is to establish limits to the amount of water that can be abstracted from a river or stream under various flow conditions. The European Union (EU) Water Framework Directive charged EU member countries with maintaining or restoring rivers to good ecological status, and one aspect of achieving good ecological status is maintaining low flows. In response, the United Kingdom introduced new restrictions on abstraction of water from surface waters and **groundwater** through an aggressive use of licenses to water users. The flow duration curve is used to define environmental flow indicators. These do not represent targets nor do they define thresholds below which no abstractions are allowed. Rather, permitted withdrawals from the river are limited when low flows occur. Under the Catchment Management Strategy, water withdrawals from rivers are permitted with stipulations about what fraction of the flow can be abstracted, paying special attention to environmental LFRs. For streams and rivers that are highly sensitive to changes in low flow, abstractions are limited to 10% of the flow when discharges are below the 95th percentile on the flow duration curve (UK Environment Agency, 2013). By 2018 the Department for Environment Food & Rural Affairs (DEFRA) reported that progress had been made such that 82% of surface water bodies and 72% of **groundwater** bodies were judged to have enough water to protect the environment.

Rivers of the world have been greatly modified by the construction of a large number of dams (see Figure 2.11). Reservoirs are constructed for a variety of purposes, but their function is to store water and release it according to a schedule, the operating rules of the reservoir. Most often, reservoir operations lead to a larger number of low-flow days than would have occurred in the unregulated river. The change in the flow duration curve can be considered an *ecodeficit* (Vogel et al., 2007), where management decisions involve trade-offs between release of water to meet environmental LFRs and release of water for human uses. Water resource managers must balance various needs when developing the operating rules for a reservoir and attempt to maximize benefits while minimizing negative impacts on **ecosystem services**.

Compared to surface water ecosystems, the status of management approaches to protect GDEs is at an early stage, and most of the development of approaches are underway in just a few countries to date (Rohde et al., 2017). Australia has been among the leaders in determining how to ensure GDEs are protected. In 2004, the National Water Initiative was instituted, under which Australian states agreed to develop **groundwater** management plans. For example, the state of New South Wales established an ecosystems policy as a component of **groundwater** policy.

New South Wales and other states in Australia have moved ahead with water sharing plans that account for the protection of GDEs. **Groundwater** extractions should be managed within the sustainable yield of **aquifer** systems, so that the ecological processes and biodiversity of their dependent ecosystems are maintained or restored. Management may involve establishment of threshold levels that are critical for ecosystem health, and controls on extraction in the proximity of **groundwater** dependent ecosystems (New South Wales Government, 2002).

Water allocation management for GDEs has not reached as mature a stage as that for LFRs in streams and rivers because of the complexity of the coupled groundwater–surface water system. It is difficult to identify and map GDEs; it is difficult to specify exactly what **groundwater** use restrictions are needed to protect the ecosystems; and it is difficult to quantify how the **groundwater** system will react to pumping with enough precision. In Australia, for example, the uncertainty about how a pumping permit to a coal mining company might harm (or not harm) a high-value GDE has generated controversy (Currell et al., 2017).

8.5 Concluding Remarks

Water required to maintain or restore healthy **ecosystem services** competes with human abstraction of water for many purposes, notably to produce food and provide energy (Perrone & Hornberger, 2014). The human appropriation of fresh water increased tremendously over the past century due to population and economic growth and the concomitant increases in demand for agricultural and industrial products and for energy in various forms. Over the past 50 years, freshwater biodiversity has decreased by more than 50% (McRae et al., 2017). Projections to the middle of the twenty-first century indicate that population will reach about 9 billion and food demand may double (Chapter 9). Given that humans already appropriate a significant fraction of the total renewable freshwater resources, it is clear that the preservation of **ecosystem services** will require insightful management of water use.

8.6 Key Points

- Categories of **ecosystem services** include provisioning services, regulating services, habitat services, and cultural services. (Section 8.1)
- Humans cause stresses in aquatic ecosystems by changing water flow, by degrading water quality, and by changing the physical landscape. (Section 8.2)
- Modern water resources management requires consideration of valuable services provided by aquatic ecosystems. (Section 8.3)
- River ecosystems are vulnerable to low flows. Water managers must consider environmental low-flow requirements (LFRs). (Section 8.4)
- Analysis of the flow duration curve for a river can provide indices to guide management decisions to balance environmental LFRs with other uses of water. (Section 8.4)
- Approaches to manage groundwater withdrawals to balance the requirements of **groundwater-dependent ecosystems** (GDEs) with other uses of water are being developed in several countries. (Section 8.4)

8.7 Example Problems

Problem 8.1. There are many values added by protecting the Catskill and Delaware watersheds that provide the water supply for New York City, including recreation, fishing, and cultural amenities in the upstream watersheds. Thus, the cost savings to NYC to avoid filtration is only one piece of the valuation of **ecosystem services**. Nevertheless, it is one part that provides context to allow us to gain perspective about **ecosystem services**. One estimate is that NYC will save about US $8 billion per year just in avoided construction costs for filtration plants.

a. The NYC water system serves about 9.5 million people. What is the value of **ecosystem services** just from this avoided construction on a per person basis?

b. Compare the figure above with the average residential cost of water. The cost of residential water in NYC is about $0.75 per 100 liters, and the average residential use is about 245 liters per day. What is the annual cost of water per person?

Problem 8.2. Spring Creek flows through a mainly agricultural area in southwest Georgia in the United States. Before 1970, there was little irrigation in the area, but in the 1970s center pivot irrigation systems proliferated. The flows in Spring Creek during the summer growing season, a time of naturally low flows, dropped substantially (see Figure 8.3).

a. On average, how many days per year before 1970 was discharge in Spring Creek less than 10 cumecs? And how many for the post-1970 period?

b. If an aquatic species in Spring Creek is critically sensitive to flows below 1 cumec, give a quantitative estimate of the change in the duration of time this species was stressed in years after 1982 relative to years before 1970.

8.8 Suggested Reading

Vörösmarty, C. J., McIntyre, P. B., Gessner, M. O., Dudgeon, D., Prusevich, A., Green, P., et al. (2010). Global threats to human water security and river biodiversity. *Nature, 467*(7315), 555–561.

8.9 References

Currell, M. J., Werner, A. D., McGrath, C., Webb, J. A., & Berkman M. (2017). Problems with the application of hydrogeological science to regulation of Australian mining projects: Carmichael Mine and Doongmabulla Springs. *Journal of Hydrology, 548*, 674–682.

McRae, L., Deinet, S., & Freeman, R. (2017). The diversity-weighted living planet index: Controlling for taxonomic bias in a global biodiversity indicator. *PLoS ONE, 12*(1), e0169156.

Millennium Ecosystem Assessment. (2005). *Current state and trends assessment: Fresh water*. Washington, DC: Island Press.

National Research Council. (2005). *Valuing ecosystem services: Toward better environmental decision-making*. Washington, DC: National Academies Press.

New South Wales Government, Department of Land & Water Conservation (2002). *The NSW state groundwater dependent ecosystems policy*. http://www.water.nsw.gov.au/__data/assets/pdf_file/0005/547844/groundwater_dependent_ecosystem_policy_300402.pdf

O'Donnell, E. L., & Talbot-Jones, J. (2018). Creating legal rights for rivers: Lessons from Australia, New Zealand, and India. *Ecology and Society, 23*(1), 7.

Perrone, D., & Hornberger, G. M. (2014). Water, food, and energy security: Scrambling for resources or solutions? *WIREs Water, 1*(1), 49–68.

Rohde, M. M., Froend, R., & Howard, J. (2017). A global synthesis of managing groundwater dependent ecosystems under sustainable groundwater policy. *Groundwater, 55*(3), 293–301.

Turak, E., Dudgeon, D., Harrison, I. J., Freyhof, J., De Wever, A., et al. (2017). Observations of inland water biodiversity: Progress, needs and priorities. In M. Walters & R. J. Scholes (Eds.), *The GEO handbook on biodiversity observation networks* (pp. 165–186). Cham, Switzerland: Springer International.

UK Environment Agency. (2013). Environmental Flow Indicator. http://webarchive.nationalarchives.gov.uk/20140328104910/http://cdn.environment-agency.gov.uk/LIT_7935_811630.pdf

Vogel, R. M., Sieber, J., Archfield, S. A., Smith, M. P., Apse, C. D., & Huber-Lee, A. (2007). Relations among storage, yield, and instream flow. *Water Resources Research*, *43*(5), W05403.

Vörösmarty, C. J., McIntyre, P. B., Gessner, M. O., Dudgeon, D., Prusevich, A., et al. (2010). Global threats to human water security and river biodiversity. *Nature*, *467*(7315), 555–561.

Zhang, B., Shi, Y.-T., Liu, J.-H., Xu, J., & Xie, G.-D. (2017). Economic values and dominant providers of key ecosystem services of wetlands in Beijing, China. *Ecological Indicators*, *77*, 48–58.

III

Anthropogenic Drivers of Change

CHAPTER 9

Population

9.1 Introduction

Human appropriation of freshwater resources is stretching the limits of sustainability. The expanding use and overuse of fresh water has largely been driven by the need to satisfy the demands of people. Therefore, water resources planning and management must consider how the population is expected to change in the future.

The world's population is increasing, and there is a roughly 25% chance that the population will stabilize or begin to decrease by 2100 (Gerland et al., 2014). An estimated 7.5 billion people live on Earth as of 2017 (Table 9.1) (Population Division of Department of Economic and Social Affairs at United Nations, 2017). This estimate is nearly three times the population of 1950. The most recent medium projection for the world's population by 2100 is more than 11 billion people (Figure 9.1). Different fertility and mortality rates across countries have resulted in different projections regionally for 2100 (see Table 9.1; Figure 9.2) (Population Division of Department of Economic and Social Affairs at United Nations, 2017). Population projections are not exact, of course, but we have 95% confidence that the population in 2100 will be between 9.6 and 13.2 billion people.

Although the population is projected to increase, the rate of population growth is projected to decrease from 2.6% annually in 2015 to 0.66% annually in 2100. Globally, fertility rates are decreasing and have been for more than a century. The large growth in population in the last ~100 years is linked closely to a decrease in mortality rates, especially among youth. Today's youth population is the largest in history. Although Asia has the largest population, the highest rate of growth is in Africa (see Table 9.1). Children under 15 years old represent 41% of the 2017 African population,

Table 9.1 Regional population estimates (2017) and medium projections (2100). Data from Population Division of Department of Economic and Social Affairs at United Nations 2017. Percentages may not sum to 100% due to roundoff.

Region	Area (km^2)	2017 Estimated (people)	2017 Percentage of total	2100 Projected (people)	2100 Percentage of total
Africa	33.28 million	1.26 billion	17	4.47 billion	40
Asia	46.46 million	4.50 billion	60	4.78 billion	43
Europe	97.84 million	0.74 billion	10	0.65 billion	6
Latin America and Caribbean	23.44 million	0.65 billion	9	0.71 billion	6
Northern America	106.69 million	0.36 billion	5	0.50 billion	4
Oceania	10.49 million	0.04 billion	1	0.07 billion	1
Total		7.55 billion		11.18 billion	

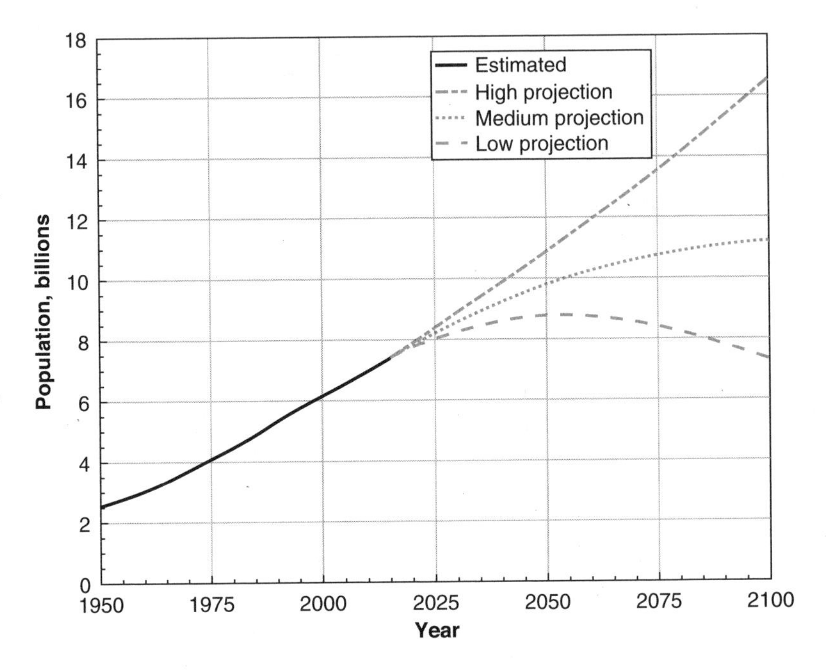

Figure 9.1 Global population projections through 2100. Although the medium projection is 11.2 billion people, there is a wide range of projections. There is 95% confidence that the population in 2100 will be between 9.6 billion and 13.2 billion people. Data from Population Division of Department of Economic and Social Affairs at United Nations 2017.

and these youth will reach childbearing age themselves in the upcoming decades. By the end of this century, 40% of the world's population will live in Africa, and 43% of the world's population will live in Asia (see Table 9.1).

Patterns of population changes in the early 2000s are remarkably different than the patterns of the 1950s, with urbanization playing a major role. As the population grows and shifts to urban areas, it is likely that more people will migrate in search of better opportunities or in response to conflicts (Population Division of Department of Economic and Social Affairs at United Nations,

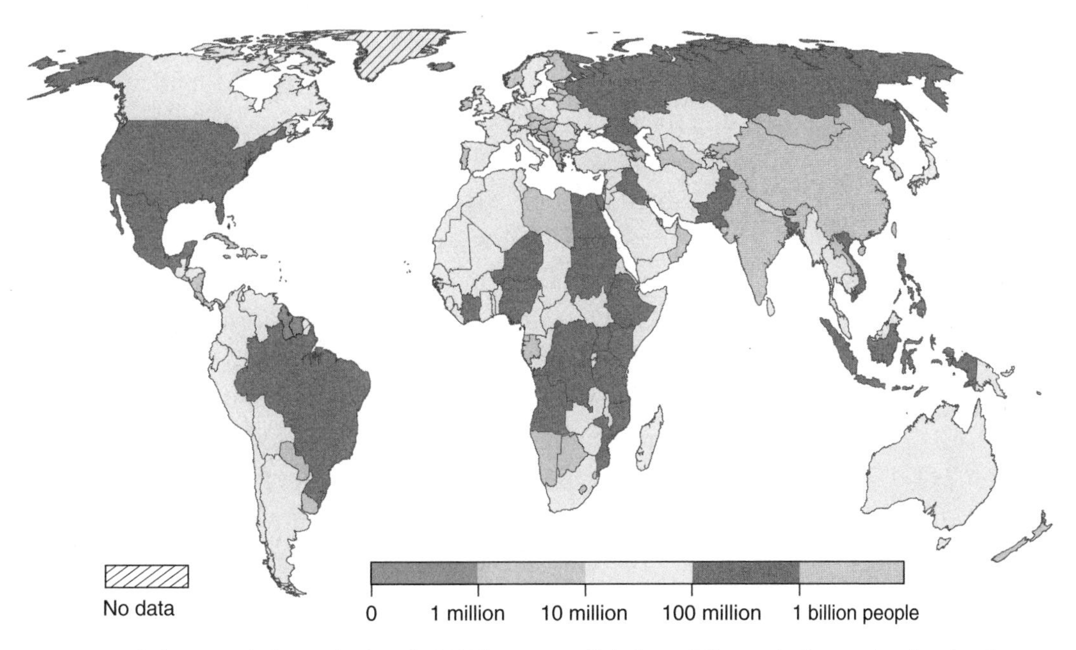

Figure 9.2 Medium population projections for 2100 by country. Global population projections and regional projections for Asia and Africa depend on a few countries with uncertain fertility rates, mortality rates, and demographic profiles. Data from Population Division of Department of Economic and Social Affairs at United Nations 2017.

2017). In the future, most people will live in megacities, with urban areas in Africa and Asia absorbing the majority of the projected population growth. Careful management of water use and investments in urban water infrastructure will be essential to accommodate the expected high rates of population growth and the growth of megacities.

9.2 Data and Models for Calculating Estimates and Projections

A key component to understanding population estimates and projections is the data and the methodology used in the calculations. The data required to make informed estimates of future population include fertility, mortality, migration, and demographic information such as age and gender. Such data are collected through official counts (enumerations) and surveys of a country's population (censuses). Because census efforts require significant investments both financially and politically, demographic and population information is more likely to be limited, lacking, or unreliable in low-income countries. This not only affects enumerations but also estimates and projections. Adult mortality and migration data are often deemed inadequate, increasing the uncertainty of population estimates and projections (Population Division of Department of Economic and Social Affairs at United Nations, 2015).

Demographic models are used to forecast future populations based on census data. The models incorporate expected birth rates, death rates, and migration rates based on historical trends and assumptions about future changes in these values. The range in global population projections (see Figure 9.1) reflects the uncertainty assigned to the estimates and the assumptions embedded in population models. Regardless of the level of uncertainty, it is clear that significant growth will continue throughout this century. This raises the question of how resources—water resources in particular—will be impacted. Will water shortages lead to unwelcome changes for the human population? Will technology and conservation measures come to the rescue? At a most basic level, we can explore some population arguments that have evolved over the past few centuries.

9.3 Population Arguments

In 1798, Malthus published *An Essay on the Principle of Population*, one component of which was the assertion that population grows geometrically while food supply grows arithmetically (Malthus, 1959). The dire implication of this assertion is that human population is bound to outstrip the food supply ultimately leading to famine and starvation. The population–food disaster, as defined by Malthus, has not come about, largely because of the Green Revolution. The Green Revolution was marked by the development of fertilizers, pesticides, high-yielding grain varieties, agricultural intensification on arable lands, and irrigation infrastructure in the early to mid-twentieth century. These technological innovations allowed the global food supply to increase faster than the growth of our global population. In short, we were able to avoid what has been dubbed a **Malthusian catastrophe** (Box 9.1).

Current arguments reflect a debate about the overall impact to be expected from population growth in the future, but often there is little *direct* discussion about the role of water, even though it is one of the most basic needs of our society.

Box 9.1 Geometric and Arithmetic Growth—the Malthusian Catastrophe

Perhaps one of the most controversial components of the first edition of Malthus's essay is the assertion that population grows geometrically while food supply grows arithmetically. The dire implication is that human population is bound to outstrip the food supply and ultimately lead to famine and starvation—the Malthusian catastrophe.

To gain a better understanding of the catastrophe, consider a very simple model of population growth, which considers only fertility rate (B) and mortality rate (D). If the initial population at time zero is N_0, the population at the next generation will be $N_1 = N_0 + (B - D)N_0 = (1 + B - D)N_0$. To make the example numerically concrete, take the initial population to be 1,000 and $B - D$ to be 0.25 (i.e., the birth rate exceeds the death rate by 25%). For a series of generations, $N_1 - N_{10}$, we then have:

$$N_1 = 1.25N_0 = 1{,}250$$
$$N_2 = 1.25N_1 = (1.25)^2N_0 = 1{,}563$$
$$N_3 = 1.25N_2 = (1.25)^3N_0 = 1{,}953$$
$$\vdots$$
$$N_{10} = 1.25N_9 = (1.25)^9N_0 = 7{,}451$$

So in 10 generations the population has increased more than sevenfold. This is geometric growth.

Next suppose that food supply, F, grows at an arithmetic rate. That is, the food supply for each generation increases by a constant amount. To be concrete, take F to be 1,000 and the increment to be 250 units.

$$F_1 = F_0 + 250 = 1{,}250$$
$$F_2 = F_1 + 250 = F_0 + 2(250) = 1{,}500$$
$$F_3 = F_2 + 250 = F_0 + 3(250) = 1{,}750$$
$$\vdots$$
$$F_{10} = F_9 + 250 = F_0 + 10(250) = 3{,}500$$

The curve for the geometric increase in population continually gets steeper whereas the curve for the arithmetic increase in food is a straight line (Figure B9.1.1). After 10 generations, the ratio of food to population has gone from 1.0 unit of food per person to 0.16 units of food per person. That is, an individual in the tenth generation would have less than 20% of the food that an individual in the original population had.

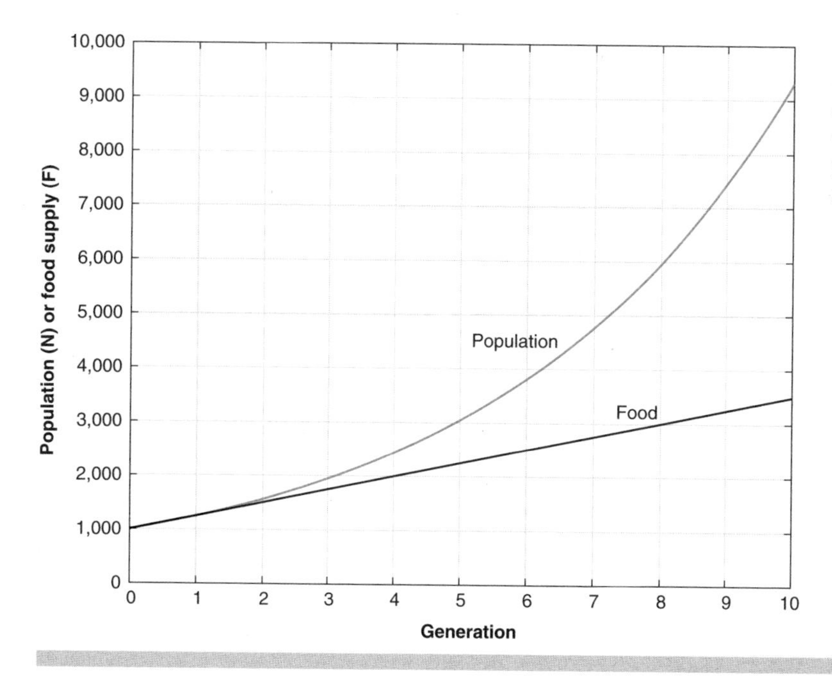

Figure B9.1.1 Geometric population growth and arithmetic food growth, where N represents human population and F represents food supply.

9.3.1 Argument: Population Growth Induces Positive Change and Innovation

One suggestion concerning population is that the larger the population, the greater will be the creation and adoption of technological innovations (Simon & Kahn, 1984). That is, more people lead to more brains, and more brains lead to more ideas. Although a higher population may lead to short-run negative impacts, the long-run positive impacts are greater (Simon, 1981). In short, this argument, promoted by Julian Simon, suggests that humans are the "ultimate resource," that technological change has been underestimated in the past, and consequently that technological change will likely be underestimated in the future.

Simon maintained that Earth's natural resource base will continue to increase throughout the millennia (Simon & Kahn, 1984). The argument with respect to water is that efficient and affordable technologies to move water from place to place, to remove pollutants so water can be reused, and to desalinate sea water will resolve any issues related to water shortages. This favorable view of unrestrained population growth is not shared widely. Some scholars believe that Simon's optimistic findings are the result of questionable model assumptions (Ahlburg, 1998).

9.3.2 Argument: Environmental Impact Is Nonlinear

The human footprint on the natural environment arises not only from the sheer number of people but also from their demand for goods and services (e.g., how much disposable income they have) and the technologies used to satisfy those demands (e.g., industrialization). The **I = PAT** equation provides a conceptually useful way to think about the environmental impact (*I*) of population (*P*), affluence (*A*), and technology (*T*) (Commoner et al., 1971). Since the conception of the **I = PAT** equation in the early 1970s, the global population has grown by nearly 4 billion people, global *gross domestic product* (GDP) has nearly quadrupled, and technological developments have provided both the cure to and the cause of environmental impacts. For instance, **groundwater** pumps allow us to access water resources previously inaccessible, but the depletion of **groundwater** has impacted ecosystems (Perkin et al., 2017).

Box 9.2 The I = PAT Equation

Many cases can be considered within the conceptual framework that links environmental impact with population, affluence, and technology. Consider the impact on water resources of supplying irrigation to crops in an arid environment. Arizona is a state in the United States where agriculture is practiced in quite arid conditions. In recent years, about 900 farms in Arizona, with an average size of 80 hectares, grew cotton. The cotton in Arizona requires about 750 mm of water to be applied to the ground surface over the growing season (see Table 5.7). The "population" in this case can be considered to be represented by the number of farms and the "affluence" by the average size of a farm. The required irrigation water reflects the technologies used to provide irrigation efficiently, so "technology" can be represented by the required irrigation water. The "impact" is then calculated to be the amount of water used to grow cotton in Arizona.

Amount of water	=	Number of farms used for cotton	×	Farm size	×	Irrigation required
I	=	*P*	×	*A*	×	*T*
(5.4×10^8 m^3 of water)	=	(900 farms)	×	(80 ha/farm)	×	(0.75 m depth of water applied per ha)

The impact in this case is the use of 540 million cubic meters of water each year to grow cotton in an arid landscape. For comparison, this is about the same amount of water supplied to 3.9 million residents of Los Angeles each year.

Although the application of the equation in cases like this where the data are available is straightforward, it is not simple to use the I = PAT equation to project what changes might occur in the future. How will irrigation technologies change in the future to lessen the water required? Might genetic engineering of crops lessen cotton's transpiration demands? Will demand for cotton continue to be high or will substitute fibers take over? Will climate change drive water requirements even higher in the future? Forecasting for the future requires assumptions to be made. Furthermore, changing conditions can be benign for some ranges then suddenly cause a large impact when some threshold is crossed.

The deceptively simple **I = PAT** equation hides considerable complexity (Box 9.2). Impacts do not necessarily arise smoothly as changes in population, affluence, or technology occur. Nonlinear relationships, including thresholds below which there is little impact and above which there may be catastrophic impact, are the norm. For example, trees and other vegetation living near streams or other wetlands require the presence of a **groundwater** table within some depth range below the ground surface. If **groundwater** pumping is controlled so the critical depth of the **water table** is maintained, the vegetation will persist; if pumping leads to lowering of the **groundwater** table below the critical depth, the vegetation will fail, with potential consequences for human populations living nearby.

Nonlinear aspects of the **I = PAT** equation do not have to lead to catastrophic impact and can be beneficial (e.g., some technologies).

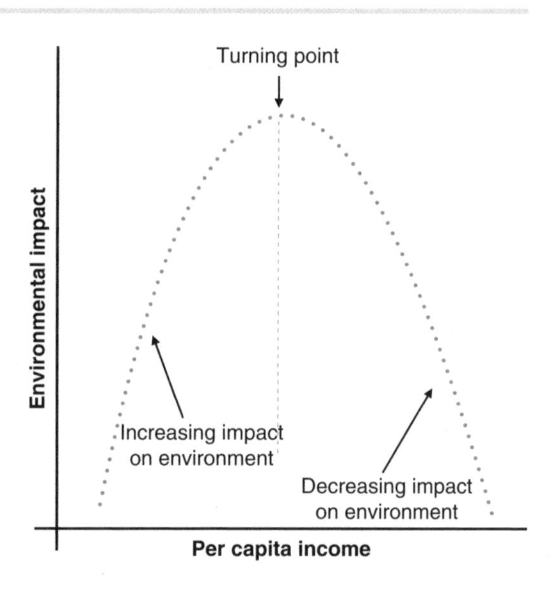

Figure 9.3 Stylized Kuznets curve. As society becomes affluent, there is more disposable income to invest in environmental friendly technologies; as a result, there can be a transition from high environmental degradation to low environmental degradation.

For example, country-level GDP and environmental impact may follow a *Kuznets curve* (Goklany, 2009). This inverted *U*-shape curve occurs (Figure 9.3) in countries where there is a transition from high environmental degradation to low environmental degradation as the society becomes affluent enough to invest in environmentally friendly technologies. Such an effect is evident for water resources in that the water quality of surface waters in low-income countries is typically much worse than in high-income countries.

9.3.3 Argument: Population Growth Destroys Biodiversity

Impacts from different kinds of stress on the environment can interact. Synergies occur when the combined impact from two or more events is greater than the impacts associated with each event individually. One of the synergistic impacts on water resources is the loss of **biodiversity**, which leads to a decline in **ecosystem services** (Chapter 8). The rate of extinction of a variety of organisms has increased significantly over the past 200 to 300 years as humans have altered their uses of land and water (Ceballos et al., 2015). Ceballos et al. (2015) have postulated that the increased pressures on vertebrate species is "related to human population size and growth, which increases consumption (especially among the rich), and economic inequity." Because the current rates of loss of **biodiversity** are unprecedented over the past several million years, scientists are dubbing the current loss as the "sixth mass extinction." The decline in **ecosystem services** due to loss of **biodiversity** is estimated to be as impactful as other environmental stressors such as climate change, acidification, and nutrient pollution (Hooper et al., 2012).

9.3.4 Argument: Growth Impacts Our Climate

Climate change will have a number of impacts on water resources, and these will be unevenly experienced in various regions of the globe (Chapter 10). **Anthropogenic** climate change is driven by emissions from burning fossil fuels (e.g., coal, and natural gas). Limiting emissions in the future is a target for sidestepping pernicious water resources impacts from climate change. Two key drivers of increased emissions from fossil-fuel use are economic growth and population growth (Edenhofer et al., 2014).

We can think about population and emissions in a number of ways, including the lifetime emissions of carbon dioxide—also known as the *carbon legacy*—of an individual. The carbon legacy of an individual is linked to lifestyle and economic factors that influence per capita emissions from adopted energy technologies (Chapter 6), as well as fertility and mortality rates. For example, Americans, on average, have higher per capita emissions and live longer than Bangladeshis, and women in the United States have fewer children, on average, than women in Bangladesh. In terms of the legacy of the per capita emissions and reproductive choices—holding the variables that influence emissions constant at 2005 numbers—an American's carbon legacy is two orders of magnitude greater than a Bangladeshi's carbon legacy (Murtaugh & Schlax, 2009). In the United States, reducing the number of offspring by one has an overwhelming impact on an individual's carbon legacy (Table 9.2); having one fewer child reduces the carbon legacy of an individual more than living an automobile-free lifestyle combined with adopting a plant-based diet (Wynes & Nicholas, 2017).

Addressing the increasing greenhouse gases in the atmosphere is critical for many reasons, one of which is the impact of warming the lower atmosphere. Warming the lower atmosphere can lead to increased intensity of the **hydrological cycle**. Exactly what the impacts will be at any location are uncertain, but broad impacts

Table 9.2 Lifetime CO_2 emissions saved by an American for various human behavioral changes or actions. Data from Murtaugh and Schlax 2009.

Action	Lifetime CO_2 saved (metric tons)
Increase car's fuel economy from 20 to 30 miles per gallon	148
Reduce miles driven from 231 to 155 per week	147
Replace single-glazed windows with energy-efficient windows	121
Replace 10 75-watt incandescent bulbs with 25 watt energy-efficient lights	36
Replace old refrigerator with energy-efficient model	19
Recycle newspaper, magazines, glass, plastic, aluminum, and steel cans	17
Reduce number of children by one—constant-emission scenario	9,441

will greatly affect water resources and our environment as well as food and energy production.

9.3.5 Argument: There Is Not Enough Food

Globally, the largest user of freshwater resources is agriculture. Whether there will be enough food to meet the needs of a growing population in the developing world will depend in large part on having enough water.

Is the food supply adequate for the current population? Chronic hunger or undernourishment affects more than 800 million people per year, and more than 98% of this undernourished population lives in developing countries (FAO, 2017). Chronic hunger and undernourishment occur when people do not have sufficient calories or nutrients to live productive lives. Data indicate, however, that there is a surplus of calories globally. If there are enough calories being produced globally, why are there millions of hungry or malnourished people in the world?

Sufficient agricultural resources need not directly translate to sufficient food resources. The maldistribution of physical, political, social, and economic resources is closely linked to food insecurity. Although the calories may be available in global stocks, access to the calories may be severely limited in areas with high poverty rates. The supply of food relies on physical and natural resources, such as land, water, and fertile soil; political and social structures, such as land tenure and labor; technology, such as plows, water pumps, and fertilizers; and economic markets. In areas with inadequate local food production, trade can be fundamental to meeting demands. Food insecurity currently exists in regions that tend to use food imports to meet local demands, where poverty rates are high, and where population growth rates are high. In impoverished areas, facilitating crop diversification, market access, and political stability is fundamental to meeting the food demands of the growing population.

What is the outlook for food security in the future as we approach 8 billion people and changing demographics? Globally, it is estimated that we will need a 25% to 75% increase above 2017 agricultural production to meet 2050 food demands (Hunter et al., 2017). Through intensified production, reduced food waste, and changes in our diets, there is room for optimism that we can meet the needs of future populations (Crist et al., 2017). Success will depend on wise use of water and land resources with attention to limiting environmental degradation.

9.4 What about Water *Directly?* Is There Enough?

Many of the population arguments explored above have some relevance to water resources and water resources management. Future population growth most certainly will provide significant challenges for water resources planning and management. The uneven distribution of water resources, affluence (or its inverse, poverty), and expected population growth indicates that the challenges will vary from country to country.

As one very simple index of water availability, we can look at the total amount of annual internal freshwater resources per person per country (Figure 9.4). What we see is that the available water resources vary significantly by country. Some countries simply do not have adequate water resources internally to supply all the demands of their population. These countries must use the global market to import **virtual water. Virtual water** is the water embedded in (i.e., used in the production of) food and other goods. Moreover, even in some countries where the physical resources are available, access to safe drinking water may be low. Globally, more than 600 million people lack access to improved drinking water (Figure 9.5). Access to safe drinking water correlates with economic prosperity, because low-income and rural areas tend to lack access to water and sanitation infrastructure.

Although safe drinking water is a primary requirement—in terms of the most basic needs over time—the water to produce food and

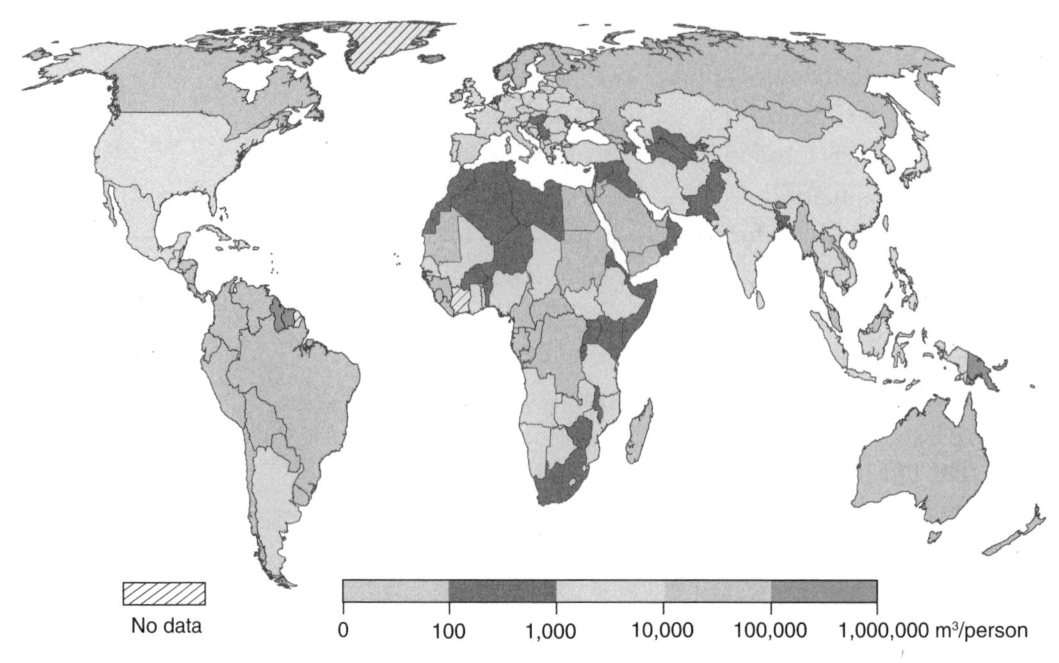

Figure 9.4 Total annual internal freshwater resources per person per country (m³/person) in the early 2010s. As population increases within a country, less internal fresh water is available per person. Data from FAO 2016.

the water used to produce electricity fall onto the list of requirements. As developing countries meet development goals, the competition for water is likely to increase, and these regions may be faced with trade-offs between using limited water resources for agricultural production or energy production. Furthermore, the benefits of environmental water flows indicate that sustaining wetlands, riparian buffers, and **groundwater**-dependent ecosystems will be critical if water resource use is to be sustainable.

Meeting the global water resource challenges as the population expands will require dealing with both the space and time variabilities of the water supply (Figure 9.6). Water demand can outpace water supply even where the water supply may be plentiful on average or across aggregate areas. Some areas with the highest expected population increases are ones where monthly and **interannual variability** in the water supply are high. To overcome the variability of the water supply over time, water must be stored in times of plenty so that it is available during dry periods. This can be accomplished by storing surface water in reservoirs or cisterns or by using the natural storage provided by **groundwater**. Of course, over the long term, more water cannot be used than is available

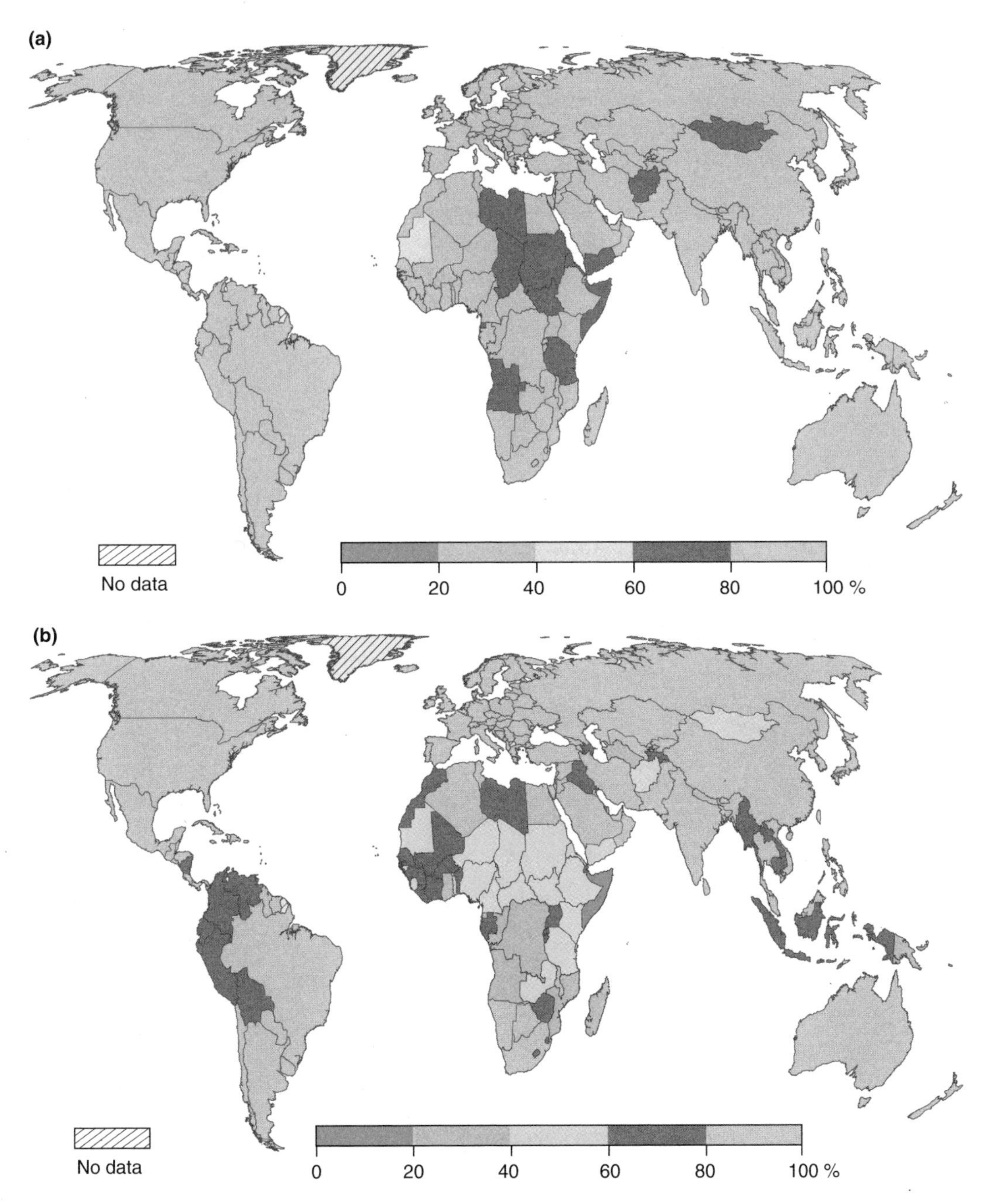

Figure 9.5 Country statistics for urban (a) and rural (b) access to safe drinking water; most recent data available shown (2001–2015, depending on country). In most countries, urban areas (a) have greater access to safe drinking water than rural areas (b). Data from FAO 2016.

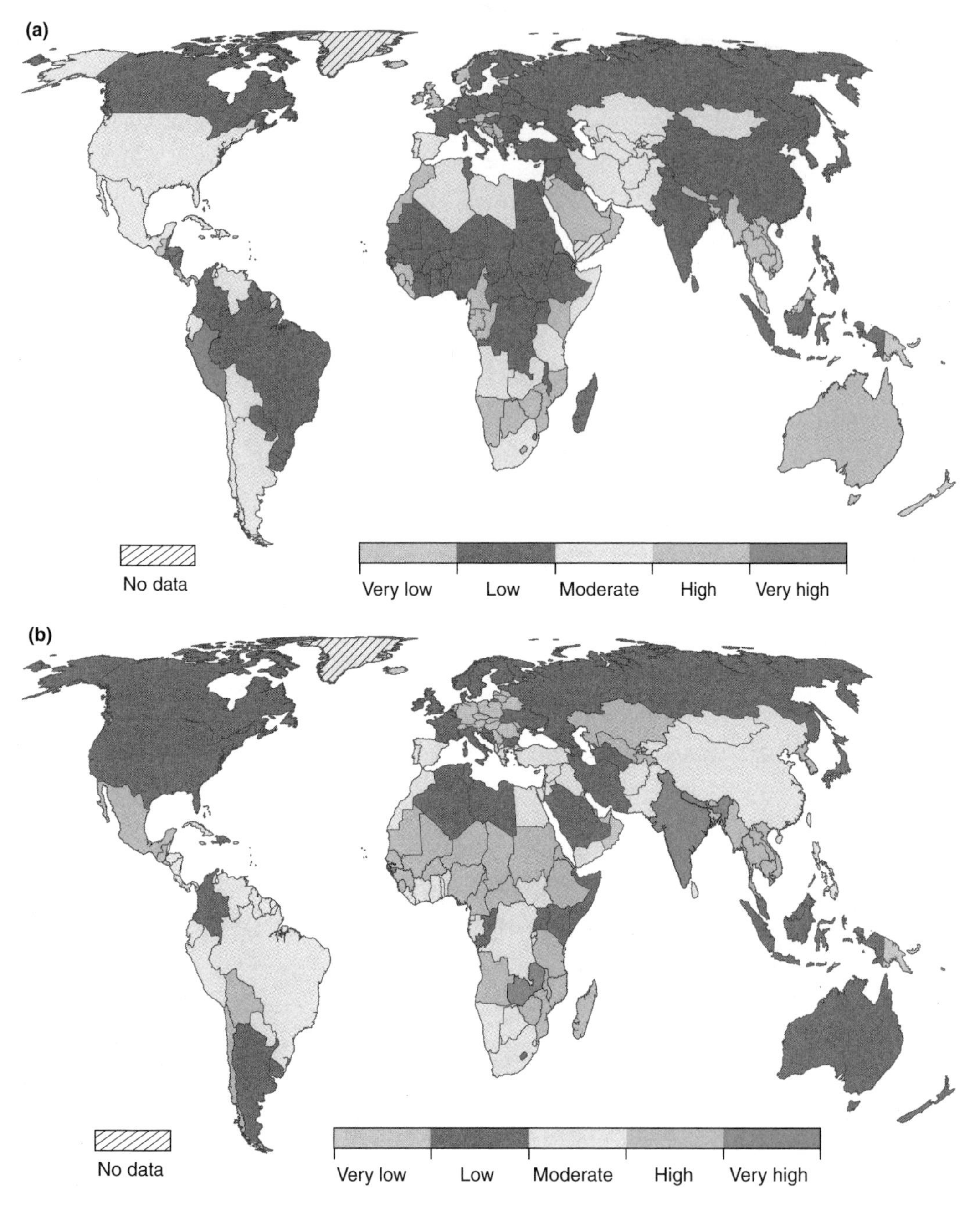

Figure 9.6 Water supply variability by country between years (a) and between months of the year (b). Data from FAO 2016.

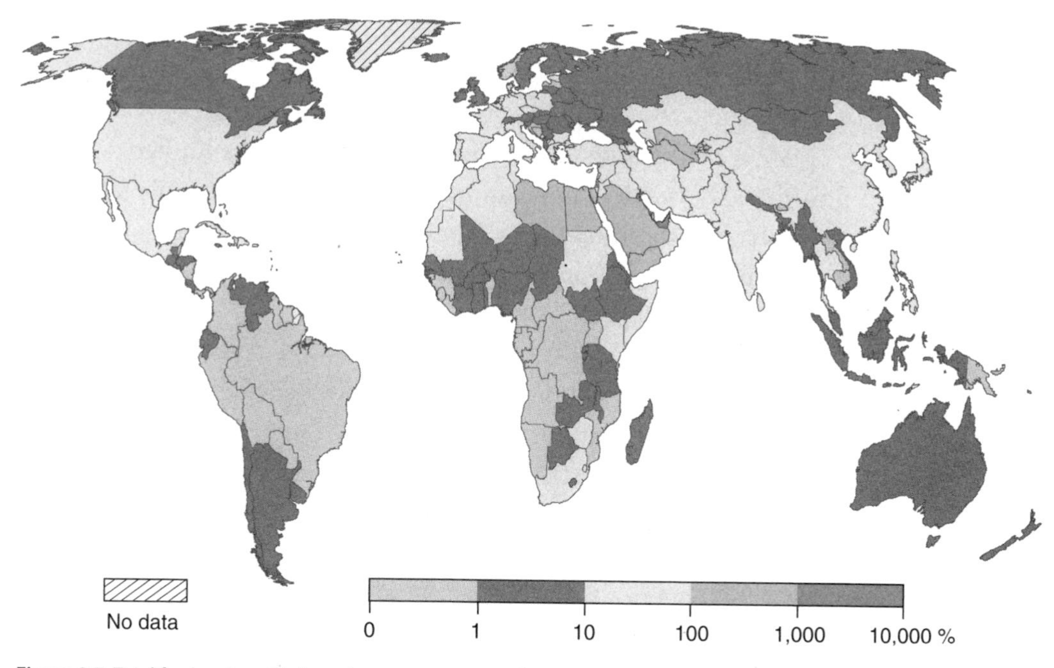

Figure 9.7 Total freshwater withdrawals as a percentage of total renewable water resources. The most recent data available are shown (2000–2016, depending on country). Data from FAO 2016.

on average. In some areas the total amount of fresh water withdrawn from surface water and **groundwater** already is larger than the total renewable fresh water (Figure 9.7), suggesting that the long-term sustainability of fresh water in these areas is at risk.

9.5 Conclusion

As we strive to deliver all the basic needs to our global population and to raise the living standards in low-income countries, we must consider the role of both technological and nontechnological solutions in supplying potable drinking water to all (Perrone & Hornberger, 2014). In short, we do not necessarily have a global water scarcity problem, but we do have spatial and temporal water distribution problems. How our spatial and temporal water scarcity problems play out will depend on how we move water management forward, how we plan for an increased population, and how we deal with development in both high- and low-income countries.

9.6 Key Points

- There is a wide range in the projected population in 2100: 9.6 to 13.2 billion people. (Section 9.1)

- Future population projections are based on estimates of fertility rates, mortality rates, and future demographics such as age profiles in a few developing countries. (Section 9.2)
- Malthus's essay is often viewed within the limited scope of his assertion that population grows geometrically while food supply grows arithmetically. The essay continues to spur significant debate about population and its role in the potential collapse of our society. (Section 9.3)
- There is a wide range of population arguments that view continued population growth as either beneficial or detrimental to humans and the environment. (Section 9.3)
- How our spatial and temporal water scarcity problems play out will depend on how we move water management forward, how we plan for an increased population, and how we plan for the development of many low-income countries. (Section 9.4)

9.7 Example Problems

Problem 9.1. Estimates for global population and total water consumption from 1960 to 2005 (Chen et al., 2016) show steady increases in each (Table 9.3). Recall that consumptive use of water is that portion that is evaporated or transpired to the atmosphere and thus is not available for direct reuse downstream (Chapter 2).

a. Over the time period 1960–2005, is population growth arithmetic or geometric?

b. Over the time period 1960–2005, is the growth in water consumption arithmetic or geometric?

Table 9.3 Global population and total water consumption from 1960 to 2005. Data from Chen et al. 2016.

Year	Population (billions)	Water consumption (thousands of km^3)
1960	3.03	2.00
1965	3.33	2.30
1970	3.69	2.60
1975	4.07	2.90
1980	4.45	3.20
1985	4.86	3.37
1990	5.32	3.55
1995	5.74	3.75
2000	6.13	3.90
2005	6.51	4.10

c. If the population in 2050 grows to 9.7 billion, estimate the water consumption at that future time.

d. An estimate of the total renewable freshwater resources globally is 39,000 km^3. What fraction of that would be *consumed* in 2050 according to your estimate from the previous question?

Problem 9.2. The agricultural GDP of the United States per person in 1960 was about $0.11 per capita and in 2000 was about $0.43 per capita (in 2000 dollars). The corresponding population of the United States in these years was 179.3 million and 281.4 million, respectively. One *impact* of agriculture on the environment is the amount of water withdrawn for irrigation. In 1960, 110 million gallons per day were withdrawn; in 2000, 128 million gallons per day were withdrawn. Note that to get the total withdrawn for the year, these values are multiplied by the number of days in the year. (Data from Dieter et al., 2018; Hobbs & Stoops, 2002)

If we apply the **I = PAT** equation for this case, the impact, *I*, is the total water withdrawn for irrigation over the year; the population, *P*, is just that; the affluence, *A*, is represented by the GDP per capita; and the technology term, *T*, accounts for any number of things that increase the efficiency of the use of water.

a. Given the values above, calculate *T*.

b. The inverse of this expression of *T* is known as water productivity, the economic value of agricultural goods per gallon of water applied, sometimes colloquially referred to as the "crop per drop." Calculate the water productivity of agriculture in 1960 and in 2000.

c. Discuss the potential reasons for changes in the values that you have just calculated for the period between 1960 and 2000.

Problem 9.3. The *water footprint* of a country is calculated as the amount of water used per capita, accounting for both direct water use in the country and the **virtual water** consumed through importing food and other products. As the affluence of the population in a developing country increases, the expectation is that the water footprint of the country will increase; however, it also might be expected that as high levels of affluence are attained, the water footprint would decrease because investments in conservation would be more likely. That is, we might expect that a Kuznets curve would describe the relationship between water footprint and a measure of affluence such as the GDP per capita. For a selection of 158 countries around the world, such a relationship is suggested (Figure 9.8).

a. Of the 158 countries shown in Figure 9.8, 139 have a GDP per capita of less than US $40,000. Many of these countries have high population growth rates and aspire to increase the economic prosperity of the country. The expectations are that the countries will navigate the

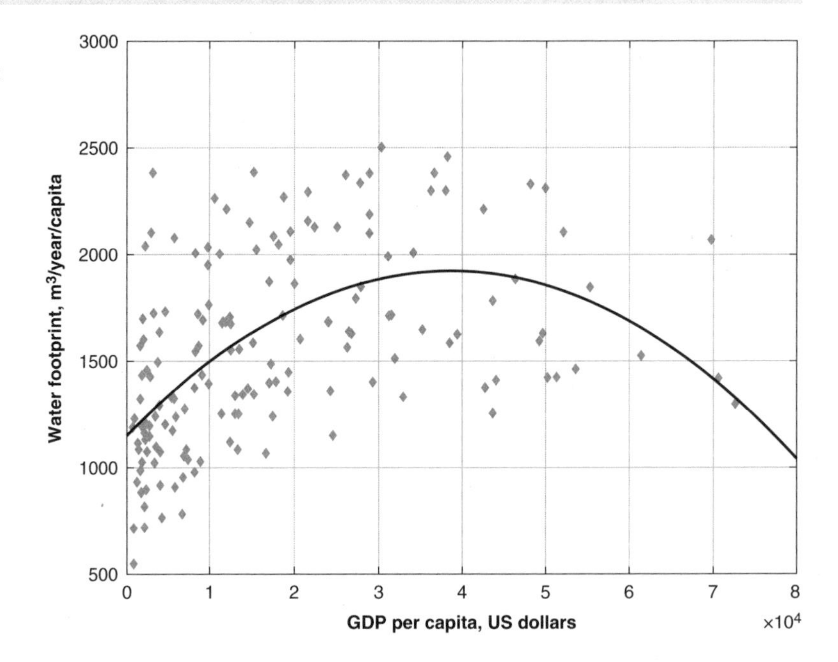

Figure 9.8 A parabola fit to data for gross domestic product per capita and water footprint per capita for 158 countries. Data from US Central Intelligence Agency 2018; Mekonnen and Hoekstra 2011.

demographic transition—that the rates of economic growth will exceed the rates of population growth, resulting in significant increases in real GDP per capita. For example, Bangladesh, which now has a per capita GDP of $4,200, is likely to see that number increase to $20,000 or so by 2050. How would you expect the overall global water footprint to change over the next three or four decades? Describe how the water footprint of Bangladesh, based on the Kuznets curve (see Figure 9.8), would change if expectations are realized.

b. The countries that currently have a GDP per capita of about $15,000 or more and that will sustain economic growth of 3% per year or more will surpass the $40,000 mark by 2050. Discuss your expectations for changes in overall water footprint later in this century based on the Kuznets curve (see Figure 9.8).

c. The Kuznets curve implicitly reflects many aspects of a country, its people, and its economy. One major country—the United States—was left off of the graph. For the United States, the GDP per capita is $59,500, and the water footprint is 2,842 m^3/year/person. Look at where the United States would plot on Figure 9.8. Does this knowledge alter your thinking relative to the previous question?

9.8. Suggested Reading

Mankin, J. S., Viviroli, D., Mekonnen, M. M., Hoekstra, A. Y., Horton, R. M., et al. (2017). Influence of internal variability on population exposure to hydroclimatic changes. *Environmental Research Letters*, *12*(4), 44007.

9.9. References

Ahlburg, D. A. (1998). Julian Simon and the population growth debate. *Population and Development Review*, *24*(2), 317–327.

Ceballos, G., Ehrlich, P. R., Barnosky, A. D., Garcia, A., Pringle, R. M., & Palmer, T. M. (2015). Accelerated modern human—induced species losses: Entering the sixth mass extinction. *Sciences Advances*, *1*(5), e1400253.

Chen, J., Shi, H., Sivakumar, B., & Peart, M. R. (2016). Population, water, food, energy and dams. *Renewable and Sustainable Energy Reviews*, *56*, 18–28.

Commoner, B., Corr, M., & Stamler, P. J. (1971). *The closing circle: Nature, man, and technology*. New York: Knopf.

Crist, E., Mora, C., & Engelman, R. (2017). The interaction of human population, food production, and biodiversity protection. *Science*, *356*(6335), 260 LP-264.

Dieter, C. A., Maupin, M. A., Caldwell, R. R., Harris, M. A., Ivahnenko, T. I., et al. (2018). *Estimated use of water in the United States in 2015*. Circular 1441. Reston, VA: US Geological Survey. https://doi.org/10.3133/cir1441

Edenhofer, O., Pichs-Madruga, R., Sokona, Y., Agrawala, S., Bashmakov, I. A., et al. (2014). *IPCC, 2014: Summary for policymakers*. In *Climate change 2014: Mitigation of climate change. Contribution of Working Group III to the Fifth Assessment Report of the Intergovernmental Panel on Climate Change*. Cambridge: Cambridge University Press. https://www.ipcc.ch/site/assets/uploads/2018/02/ipcc_wg3_ar5_summary-for-policymakers.pdf

Food and Agriculture Organization (FAO) of the United Nations. (2016). AQUASTAT Main Database. http://www.fao.org/nr/water/aquastat/data/query/index.html?lang=en

Food and Agriculture Organization (FAO) of the United Nations. (2017). *The state of food security and nutrition in the world*. http://www.fao.org/state-of-food-security-nutrition/en/

Gerland, P., Raftery, A. E., Sevcikova, H, Li, N., Gu, D., et al. (2014). World population stabilization unlikely this century. *Science*, *346*(6206), 234–237.

Goklany, I. M. (2009). Have increases in population, affluence and technology worsened human and environmental well-being. *Electronic Journal of Sustainable Development*, *1*(3), 15.

Hobbs, F., & Stoops, N. (2002). *Demographic trends in the 20th century*. Census 2000 Special Reports, CENSR-4. US Department of Commerce, US Census Bureau. https://www.census.gov/prod/2002pubs/censr-4.pdf

Hooper, D. U., Adair, E. C., Cardinale, B. J., Byrnes, J. E. K., Hungate, B. A., et al. (2012). A global synthesis reveals biodiversity loss as a major driver of ecosystem change. *Nature*, *486*(7401), 105–108.

Hunter, M. C., Smith, R. G., Schipanski, M. E., Atwood, L. W., & Mortensen, D. A. (2017). Agriculture in 2050: Recalibrating targets for sustainable intensification. *BioScience, 67*(4), 386–391.

Malthus, T. R. (1959). *Population: The first essay.* Ann Arbor: University of Michigan Press. (Original work published 1798)

Mekonnen, M. M., & Hoekstra, A. Y. (2011). *National water footprint accounts: The green, blue and gray water footprint of production and consumption.* Value of Water Research Report Series No. 50. Delft, the Netherlands. http://www.waterfootprint.org/Reports/Report50-NationalWaterFootprints-Vol1.pdf

Murtaugh, P. A., & Schlax, M. G. (2009). Reproduction and the carbon legacies of individuals. *Global Environmental Change, 19*(1), 14–20.

Perkin, J. S., Gido, K. B., Falke, J. A., Fausch, K. D., Crockett, H., et al. (2017). Groundwater declines are linked to changes in Great Plains stream fish assemblages. *Proceedings of the National Academy of Sciences of the United States of America, 114*(28), 7373–7378.

Perrone, D., & Hornberger, G. M. (2014). Water, food, and energy security: Scrambling for resources or solutions? *WIREs Water, 1*(1), 49–68.

Population Division of Department of Economic and Social Affairs at United Nations. (2015). *World population prospects: Key findings and advance tables.* https://doi.org/10.1017/CBO9781107415324.004

Population Division of Department of Economic and Social Affairs at United Nations. (2017). *World population prospects: The 2017 revision, key findings and advance tables.* No. Working Paper No. ESA/P/WP/248. https://population.un.org/wpp/Publications/Files/WPP2017_KeyFindings.pdf

Simon, J. (1981). Environmental disruption or environmental improvement? *Social Science Quarterly, 62*(1), 30–43.

Simon, J., & Kahn, H. (Eds.). (1984). *The resourceful earth: A response to Global 2000.* New York: Blackwell.

US Central Intelligence Agency. (2018). World fact book: Country comparison GDP - per capita (PPP). https://www.cia.gov/library/publications/the-world-factbook/rankorder/2004rank.html

Wynes, S., & Nicholas, K. A. (2017). The climate mitigation gap: Education and government recommendations miss the most effective individual actions. *Environmental Research Letters, 12*(7), 74024.

CHAPTER 10

Climate Change

10.1 Introduction

The climate of Earth is changing. In large part the change is due to the addition of carbon dioxide (CO_2) to the atmosphere from the burning of fossil fuels. Before the industrial revolution, the atmospheric concentration of CO_2 was about 280 parts per million (ppm), but CO_2 has increased steadily and now exceeds 400 ppm (Figure 10.1). CO_2 is a **greenhouse gas**. Greenhouse gases "trap" incoming **radiant energy** from the sun, heating the lower atmosphere. This process is similar to greenhouses, where heat is trapped inside during a sunny winter day. How might water resources be affected by climate change? To answer this question, we need to know (1) how increased CO_2 in the atmosphere has affected climate historically and how increased CO_2 in the atmosphere may in the future result in changes in climate variables, (2) how these changes in climate variables affect hydrological processes, and (3) how the hydrological changes will impact water resources.

10.2 Carbon Dioxide Concentrations and Our Future

The atmosphere is made up of a mixture of gases with nitrogen, oxygen, and argon comprising about 99% of the total. CO_2 is a trace gas making up only a small fraction of the total volume. Nevertheless, CO_2 has a profound effect on our climate because of its interaction with radiation. **Radiant energy** is distributed across a spectrum of wavelengths. The sun's radiation is primarily in the visible part of the spectrum, meaning our eyes can directly perceive this radiation. But sunlight is composed of light with different wavelengths. You probably have observed that a glass prism will decompose visible sunlight into a series of colors from violet to red. The separation of sunlight into various colors is because light with different wavelengths is refracted (or bent) differently by travel

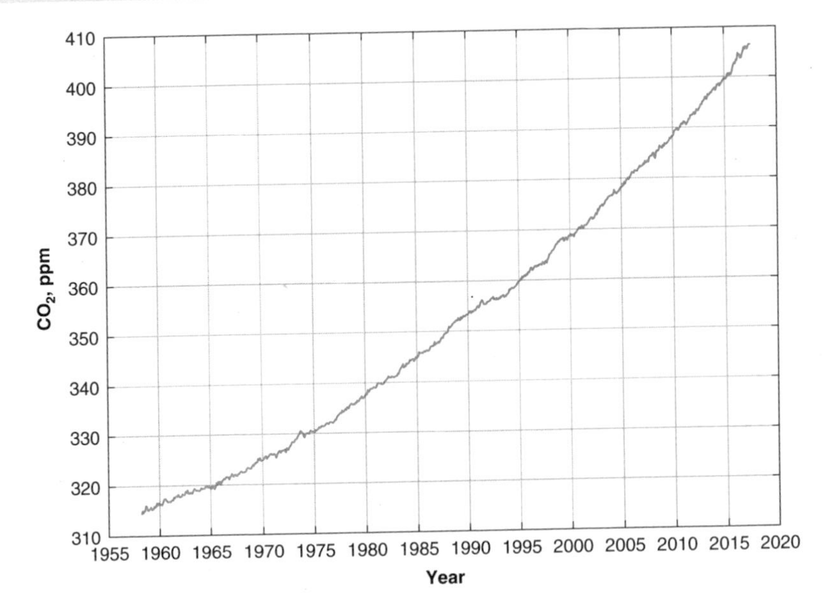

Figure 10.1 Atmospheric carbon dioxide concentration in parts per million (ppm) recorded at the Mauna Loa observatory in Hawaii. Data from Keeling et al. 2001.

through a prism. Violet is a short wavelength, and red is a long wavelength in visible radiation.

Parts of the radiation spectrum outside the visible range are not sensed directly by our eyes. Beyond the violet end of the visible spectrum are things like X-rays, which have wavelengths shorter than violet. Wavelengths longer than the red end of the visible spectrum are in the infrared range. You may have experienced or seen in movies night-vision goggles that allow a person to "see" the **infrared radiation** being emitted. For example, with the goggles, we can see the **infrared radiation** of a warm animal on a cool night. The goggles amplify the incoming radiation (including **infrared radiation**) and convert it to a phosphorescent visible image. Note that the implication is that all materials radiate, but on Earth much of the radiation is in the infrared range, which has significant implications for CO_2.

What happens to radiation when it encounters material? There are three main options: it can be transmitted through the material, it can be reflected by the material, or it can be absorbed by the material. If it is absorbed by the material, the material is heated. If you are out in the sun on a warm day, light-colored clothing will reflect much of the sunlight whereas dark-colored clothing will absorb the sun's radiation. If you wear dark-colored clothing, you will feel the warmth because the absorbed visible radiation is converted to heat. Whether **radiant energy** is transmitted, reflected, or absorbed depends on the wavelength of the radiation.

The atmosphere is largely transparent to the visible radiation of sunlight. Some of the radiation is reflected above the surface of Earth, mainly by clouds, but otherwise most of the energy reaches the surface. The average reflection coefficient of Earth as a whole, Earth's **albedo**, is about 0.3. This means that, on average, 30% of the of the incoming radiation is reflected by clouds, rocks, snow and ice, vegetation, or other materials. The remaining 70% of the radiation is absorbed and warms the absorbing materials. The absorbed energy is reradiated, but as noted above, the radiation is in the infrared part of the spectrum. This is where CO_2 becomes important. CO_2 strongly absorbs **infrared radiation**, thereby being heated. As CO_2 concentrations in the atmosphere increase, more radiation is absorbed. Consequently, the air is heated more, leading to an overall increase in air temperatures in the lower atmosphere.

10.3 Hydrological Implications of Climate Change

As noted in Chapter 1, the **hydrological cycle** is driven by energy input from the sun. Thus, any changes in the balance of incoming solar radiation, or the energy reradiated to space by Earth, implies that there will be changes in the **hydrological cycle**. These changes can be local, regional, or global. For example, the change could be due to increased temperature in certain places such as the Arctic, or large-scale atmospheric motions can be affected, changing the timing and location of monsoon circulation.

One of the main processes of the **hydrological cycle** that is affected directly by temperature is precipitation. The **Clausius-Clapeyron relationship**, an equation based on fundamental physical principles, indicates that the amount of water vapor that the atmosphere can hold increases by about 7% for every degree Celsius increase in temperature. The resulting increase in atmospheric water vapor presents what is called a positive feedback: water vapor is a very strong absorber of **infrared radiation**, so it reinforces the greenhouse effect, which leads to further temperature increase. At the same time, increases in cloudiness that can result from increases in water vapor will affect Earth's **albedo**. That is, clouds decrease the amount of solar radiation reaching the land

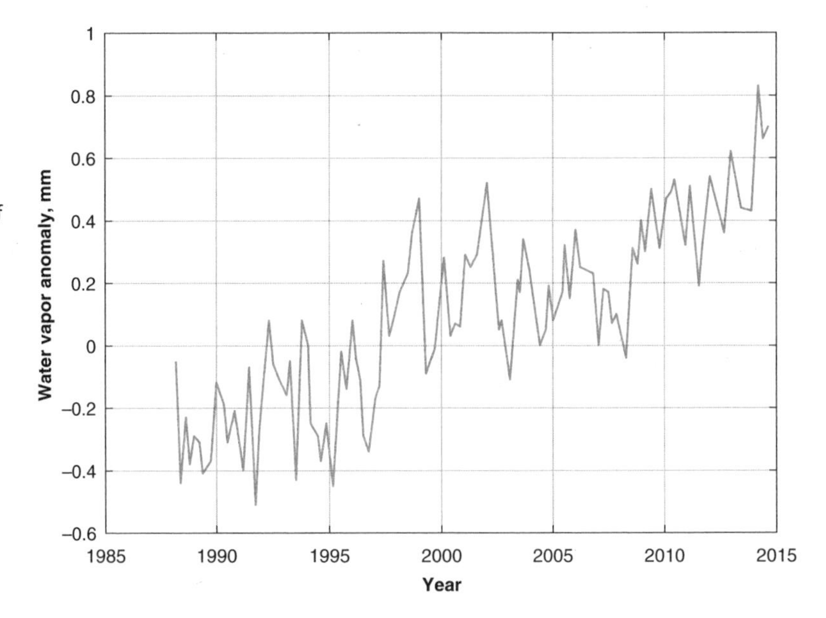

Figure 10.2 Global oceanic total column water vapor from satellite measurements corrected for El Niño and volcanic eruptions. Values expressed as millimeters of water anomaly, which is the difference between the measured water vapor and the average over the period of record. Data from Adler et al. 2017.

surface. In short, the overall effect of increases in water vapor in the atmosphere is complicated, and it is unclear how significant the positive feedback of warming caused by increases in CO_2 will become.

Measurements show that the total water vapor in the atmosphere has increased as temperatures have increased steadily over the past five decades. Where moisture is readily available—over oceans and over land areas with a humid climate—this is especially the case (Figure 10.2). The increases in water vapor are generally consistent with the **Clausius-Clapeyron relationship**. Additional water vapor in the atmosphere means that storms have more water available to feed them. The expectation of hydrologists and climate scientists is that climate change has led to an acceleration or intensification of the **hydrological cycle**. This acceleration is expected to continue as further warming occurs. Changes in the timing and spatial distribution of precipitation can be expected as the **hydrological cycle** intensifies.

More water vapor in the atmosphere should increase the chances that extreme rainfall will occur. Studies across the globe have indicated that intense precipitation has increased in frequency in the past several decades and that more of the total precipitation is occurring in periods of intense precipitation with less of the total occurring in lower intensity storms. For example, the fraction of

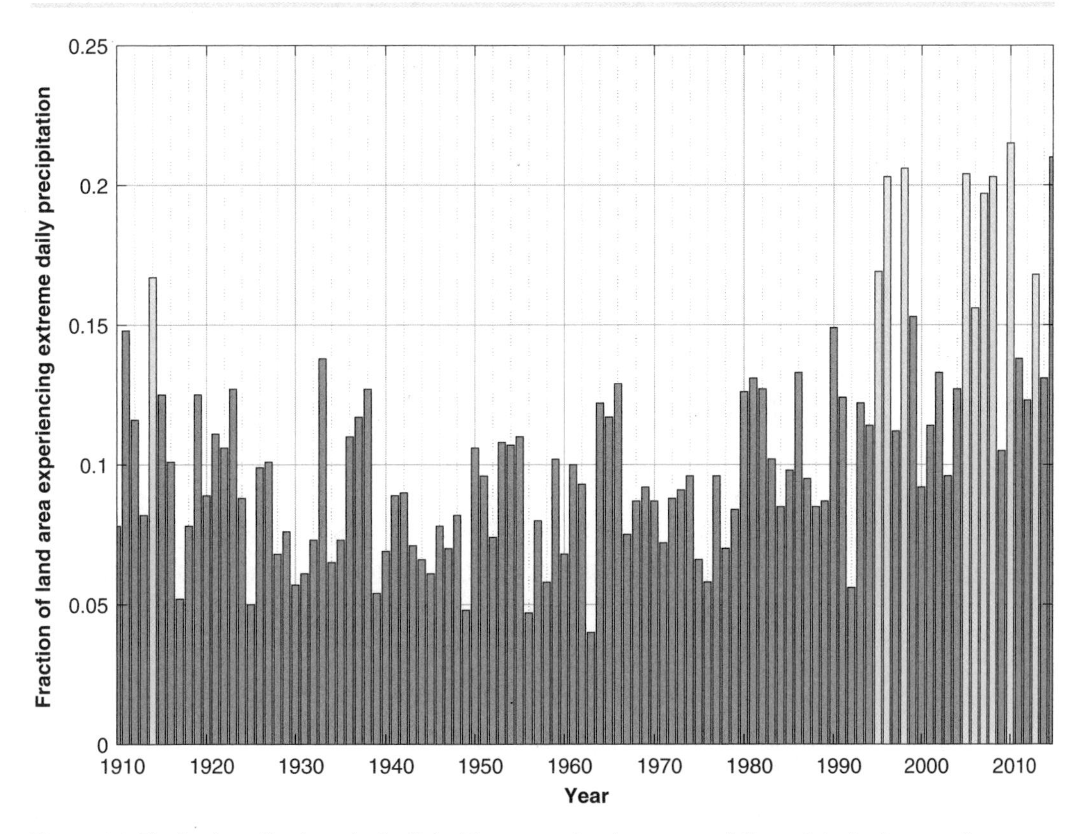

Figure 10.3 The fraction of land area in the United States experiencing extreme daily precipitation in a year. Extreme precipitation is defined as the highest tenth percentile of all recorded values. Nine of the 10 highest values (light blue) in the record occurred since 1995. Data from US Environmental Protection Agency 2016.

land area in the United States that experiences daily precipitation in the upper tenth percentile of all recorded values has increased in the last few decades (Figure 10.3). On average, we would expect that 10% of the area would have daily precipitation values in the upper tenth percentile. From 1910 to 1990, the fractional area shown in Figure 10.3 averaged a little over 9%, consistent with the expectation. After 1990, however, the average increased to almost 15%. Analyses of data for precipitation over land in "wet" and "dry" regions indicates that intense rainfall has increased regardless of the region (Donat et al., 2016). This is surprising in dry regions because atmospheric water vapor content is not expected to increase, because there is limited water available to evaporate. Computer projections indicate that intense hourly precipitation in the United States might increase by as much as 400% in the coming decades (Prein et al., 2016).

The intensification of the **hydrological cycle** will affect **evapotranspiration** as well as precipitation. The impacts of a combination of increased CO_2 and increased temperature on **evapotranspiration** are complex. Plants balance water loss through leaves with intake of CO_2 to fuel photosynthesis, so all other things being equal, plants require less water at higher CO_2 concentrations. Higher levels of CO_2, however, may increase plant growth and change leaf area, thereby increasing transpiration. Changes in temperature and precipitation patterns also impact **evapotranspiration**. The overall impact of climate change on transpiration is uncertain.

Droughts result in response to large-scale atmospheric circulations such as El Niño as well as to changes in the evaporative demand of the atmosphere. *El Niño* refers to a climate phenomenon in which unusually warm surface water in the Pacific Ocean flows eastward and results in large-scale changes in atmospheric circulation and rainfall patterns. Our understanding of how climate change will alter these circulation patterns is not clear. Nevertheless, even if the frequency of natural drought events does not change, higher temperatures are likely to cause the initiation of droughts to occur earlier, their duration to be extended, and their intensity to be greater. The Colorado River Basin in the United States is one example of how hydrological impacts may play out. The record drought that began in 2000 in the basin is attributed to record high temperatures as well as a lack of precipitation; estimates are that for every degree Celsius increase in temperature, the flow in the Colorado River is expected to decrease by about 6.5% (Udall & Overpeck, 2017). The potential hydrological consequences for the future are very large if temperatures continue to increase in the future.

10.4 Water Resources Implications of Climate Change

Major needs for freshwater resources occur in various sectors, including agriculture, energy, industry, households, and ecosystems. To meet the demands for water, the resource must be available in the right quantity at all times when needed—not too little, but also not too much—and with good quality. Changes in the **hydrological cycle** will not be uniform across the world, but the effects will include changes in river runoff, both amount and timing, changes in **groundwater** recharge, and changes in water demand in all sectors. Thus, the potential impacts on availability and use of water resources are significant.

Climate change may impact river runoff through changes in the overall amount of flow as well as the variability of flow from year to year and season to season. As noted above, runoff is expected to decrease in the Colorado River in the United States as temperatures increase (Udall & Overpeck, 2017). In contrast, the average runoff in the Nile River is expected to increase. Such changes have profound effects on water resource management, in large part because they affect the amount of water available to be stored in dams. While the average runoff in the Nile River is expected to increase, the **interannual variability** is expected to increase as well (Siam & Eltahir, 2017).

Dams and their reservoirs are planned and designed to store water in relatively wet seasons or years so it can be used during relatively dry seasons or years. Of course, the total amount of water that can be stored is limited by the amount of water available. For rivers like the Colorado, the number of years where there is failure to be able to supply demands will increase under climate change. For rivers like the Nile, it may be possible to design dams that could take advantage of increased average flows and mitigate the increased variability in flow.

Climate change can also affect seasonal variability of river flows. Precipitation patterns can shift in time, resulting in changes in the timing of runoff. Temperature increases result in earlier melting of snow in mountain watersheds, which brings earlier flow increases in the spring. In Europe, melting has led to earlier spring runoff peaks in much of the northeast whereas the changing timing of storms has led to later spring runoff peaks in some Mediterranean regions (Blöschl et al., 2017). Water resources managers must account for such changes.

Other water resources impacts will occur under changing climate. Irrigation water requirements will increase, at least for certain crops. If water availability for irrigation becomes limited, this change can have serious implications for food availability (Chapter 5). Increased temperatures can impact water quality (Chapter 12). Increasing water temperatures can limit the use of water for cooling at **thermoelectric** power plants, with a concomitant impact on energy supply (Chapter 6). Increases in water temperatures and decreases in low flow will lead to increased concentrations of contaminants, affecting aquatic ecosystems (Chapter 8).

It is important to recognize that climate change alone is not uniquely a determinant of water availability and water demand. Rather, climate change is an aggravating factor in adapting to large increases in water demand by humans in response to pressures induced by the combination of population growth and changing lifestyles (Chapter 9). The widespread appropriation of water by humans through changes of the land surface have a dominant effect on water resources across the world (Jaramillo & Destouni, 2015).

10.5 Concluding Remarks

The expected impacts of climate change on water resources are many and varied. Climate changes will not be uniform across the globe. In some regions runoff and **groundwater** recharge will decline for an average year, indicating that the total water resources available on a long-term basis will be reduced. In other regions, the reverse will be true. The water available in an average year is not the sole issue for water resource use—the year-to-year variability is important. In regions where droughts and floods occur more frequently in the future than in the past, adaptation to climate change will have to include alteration in how water resources are managed. Climate change is but one **anthropogenic** driver of change in the **hydrological cycle**. Water resources planning also will have to be done with full recognition of other drivers that have large effects on water demand and use, such as population growth, increasing urbanization, and changing energy and food demands.

10.6 Key Points

- The atmospheric concentration of CO_2 has increased from about 280 parts per million (ppm) early in the twentieth century to more than 400 ppm in 2015. (Section 10.1)
- Atmospheric CO_2 is a strong absorber of **radiant energy** from the surface of Earth, effectively trapping energy and leading to increased temperatures in the atmosphere. (Section 10.2)
- The amount of water vapor that the atmosphere can hold increases by about 7% for every degree Celsius increase in temperature

(**Clausius-Clapeyron relationship**). Thus, as air temperatures increase, the **hydrological cycle** will intensify. (Section 10.3)

- Climate change will lead to changes in total runoff from rivers as well as to shifts in seasonal flows. Extreme rainfall events and prolonged droughts will become more common. (Section 10.3)
- Planning for the use of water resources in the future will have to take the impacts of climate change into account. (Section 10.4)

10.7 Example Problems

Problem 10.1. For London in the United Kingdom, the rainfall intensity for a 1-hour event with an exceedance probability of 0.01 is 39 mm/hour based on historical data. (Recall from Section 2.4 that this means that rainfall of that intensity or greater is expected to occur on average once every 100 years.) If extreme rainfall scales according to the **Clausius-Clapeyron relationship**, how would the intensity of the 100-year, 1-hour rainfall event change for a 2°C increase in temperature in the future? An alternate scenario for the future indicates a possible temperature increase of 4°C; what would the rainfall intensity be in that case?

Problem 10.2. Water use of the Colorado River in the United States is apportioned among Western states according to a 1922 compact, with each state assigned a share (Table 10.1). By the early part of the twenty-first century, water use had equaled or exceeded the supply available from the river, and essentially nothing was left downstream of the last dam on the river (see Figure 2.12). Take the total supply over the past several decades as 15 million acre-feet per year. (An acre-foot is a US volume measure equal to a foot of water covering an acre of ground area, so 15 million acre-feet is about 18.5 km^3 per year.) Each of the states uses the assigned water for slightly different purposes, but each state derives significant economic benefit from the water.

a. Assuming that 15 million acre-feet of Colorado River water was supplied in 2014, what is the economic value per water volume supplied to each state in dollars per acre-foot?

Table 10.1 Colorado River (United States) water apportioned among Western states according to 1922 compact. Data from James et al. 2014.

State	Percentage of Colorado River water appropriated	2014 Economic value of Colorado River water (US billion $)
Arizona	18.6	185
California	29.4	657
Colorado	25.9	189
Nevada	2.0	115
New Mexico	5.6	60
Utah	11.5	70
Wyoming	7.0	22

b. If there were a 6.5% decrease in water available as projected in the future due to climate change, calculate the economic impact in 2014 dollars?

c. The 1922 compact divides the region into upper-basin states and lower-basin states. The lower-basin states are Arizona, California, and Nevada. Repeat the previous question assuming that the full 6.5% decrease is taken only from the shares of the lower-basin states.

10.8 Suggested Readings

Döll, P., et al. (2014). Integrating risks of climate change into water management. *Hydrological Sciences Journal, 60*(1), 4–13.

Vörösmarty, C. J., Green, P., Salisbury, J., & Lammers, R. B. (2000). Global water resources: Vulnerability from climate change and population growth. *Science, 289*(5477), 284–288.

10.9 References

Adler, R. F., Gu, G., Sapiano, M., Wang, J.-J., & Huffman, G. J. (2017). Global precipitation: Means, variations and trends during the Satellite Era (1979–2014). *Surveys in Geophysics, 38*(4), 679–699.

Blöschl, G., Hall, J., Parajka, J., Perdigao, R.A.P., Merz, B., et al. (2017). Changing climate shifts timing of European floods. *Science, 357*(6351), 588–590.

Donat, M. G., Lowry, A. L., Alexander, L. V., O'Gorman, P. A., & Maher, N. (2016). More extreme precipitation in the world's dry and wet regions. *Nature Climate Change, 6*, 1–7.

James, T., Evans, A., Madly, E., & Kelly, C. (2014). *The economic importance of the Colorado River to the basin region.* Tempe: W. P. Carey School of Business, Arizona State University. http://azsmart-dev.wpcarey.asu.edu/wp-content/uploads/2015/01/PTF-Final-121814.pdf

Jaramillo, F., & Destouni, G. (2015). Local flow regulation and irrigation raise global human water consumption and footprint. *Science, 350*(6265), 1248–1251.

Keeling, C. D., Piper, S. C., Bacastow, R. B., Wahlen, M., Timothy, P., et al. (2001). *Exchanges of atmospheric CO2 and 13CO2 with the terrestrial biosphere and oceans from 1978 to 2000. I. Global aspects.* SIO Reference Series, No. 01-06. University of California–San Diego. https://escholarship.org/uc/item/09v319r9#

Prein, A. F., Rasmussen, R. M., Ikeda, K., Liu, C., Clark, M. P., & Holland, G. J. (2016). The future intensification of hourly precipitation extremes. *Nature Climate Change, 7*(1), 48–52.

Siam, M. S., & Eltahir, E. A. B. (2017). Climate change enhances interannual variability of the Nile River flow. *Nature Climate Change, 7*(5), 350–354.

Udall, B., & Overpeck, J. (2017). The twenty-first century Colorado River hot drought and implications for the future. *Water Resources Research*, *53*(3), 2404–2418.

US Environmental Protection Agency (EPA). (2016). *Climate change indicators in the United States 2016*. https://www.epa.gov/climate-indicators

CHAPTER 11

Water Law

11.1 Introduction

As we move into a generation with significant growth in population and development, setting rules, both formal and informal, can be useful to promote sustainable water management. There are many useful ways to set rules, but laws and **regulations** often stand out as a formal approach. The goal of law is to prevent disputes or resolve conflicts by identifying standards and expectations formally that are enforceable by penalties or other mandatory orders against violators.

The term "water law" is part of a broader portfolio of law that encompasses elements of water quantity *and* water quality (Box 11.1). As we learned in previous chapters, water quantity and water quality are intertwined, especially when water is scarce. Many urban systems dilute water that has contaminants with larger quantities of water that do not have contaminants. Water quality, in turn, can impact the amount of water that is useable for the environment and society.

11.2 Legal Frameworks in the United States

Water policy is the sum total of **regulations**, laws, and other institutions that provide a framework for **water management**. Water policy in the United States is complex. One reason for this complexity is that various aspects of water are governed differently. Navigation and water quality are a matter of federal law and state law; many states have their own water quality **statutes** and **regulations** that either implement or expand upon the federal Clean Water Act. The federal government plays a more limited role in water quantity. Allocation of water rights (see Box 11.1) is primarily governed by state and local law. Therefore, understanding the hierarchy of legal power, as well as the various sources of law, is critical to understanding water law in the United States (Box 11.2).

Box 11.1 Water Rights versus the Right to Water

In 2010, the UN's General Assembly passed Resolution 64/292. This resolution acknowledges formally the human right to water and sanitation. Nevertheless, few countries recognize the human right to water in a legal context. South Africa is most noted for its recognition of the human right to water. Article 27 of the Constitution of the Republic of South Africa states, "Everyone has the right to have access to . . . sufficient food and water . . . The state must take reasonable legislative and other measures, within its available resources, to achieve the progressive realization of each of these rights." Even where the human right to water exists, the right may be simply aspirational (e.g., California in the United States) and not necessarily legally actionable.

Water rights are not the same as the right to water or the right of access to water. Water rights are expectations or rules about water allocation. These rights deal with water that is used in situ or water that is withdrawn from its source. Rights can be formalized expectations or customary expectations regarding the allocation of water resources.

Box 11.2 Useful Legal Definitions for the United States

The first step to understanding water law is to understand legal terms used in law.

- *Appropriations*: the withdrawal of water for various uses
- *Arbiter*: a person making a ruling or judgment (e.g., a judge or a government agency official)
- *Adjudication*: a court or agency decision about a disputed matter
- *Common law*: judge-made law derived from custom and judicial precedent rather than statutes
- *Customary water rights*: customary water use that has been formally or informally recognized with contemporary water rights
- *Enacted*: passed
- *Eminent domain*: government's right to take private property for public use by providing compensation
- *Legal doctrine*: set of rules often established through precedent in cases or case law
- *Legislation*: a law enacted
- *Ordinances*: laws enacted by local governments
- *Plaintiff*: a person or entity initiating a civil (noncriminal) lawsuit in court; an accuser
- *Preempt*: override
- *Regulations*: rules set by a government agency to implement legislation that have the force of law
- *Statutes*: laws enacted by legislative branch of government
- *Takings*: private property taken for public use (requires compensation)

11.2.1 Structure of Legal Frameworks

Legal power is distributed hierarchically among federal, state, and local legal regimes. The US Constitution prevails over other sources of law. The US Constitution assigns different powers to the three branches of the federal government, and it defines the balance of power between federal and state governments. Powers not granted to the federal government in the US Constitution are reserved for state governments. State constitutions and **statutes** establish and define the authority of local governments. Federal laws preempt conflicting state or local laws, and state law, generally, preempts conflicting local law.

Federal and state governments are structured similarly. There are three branches of federal and state government, each of which plays a very distinctive and important role in law. The legislative branch creates laws, the executive branch executes laws, and the judicial branch interprets laws and determines whether laws violate the Constitution. In the federal government, the legislative branch is referred to as Congress, and it is composed of the Senate and the House of Representatives. The legislative body of the 50 states is referred to as the state legislature, general assembly, general court, or legislative assembly depending on the state. At the federal level, executive power is vested in the president; at the state level, executive power is vested in the governor. At the federal level, the judicial branch is composed of the courts, and the Supreme Court has primacy. Local governments are structured differently than federal and state governments, and power must be granted by the state.

11.2.2 Sources of Law

The US Constitution is the supreme law of the United States, and it is the source of all governmental powers and limitations. In addition to the Constitution, there are four other sources of law (Goldfarb, 1991).

1. Statutory law. Statutory law is written and enacted by the legislative branch. These laws are referred to as **statutes** or acts. In certain cases, Congress may enact a statute to create (or eliminate) an agency. A statute provides a regulatory framework with information about how authority will be delegated and the basic standards that must be followed, but a statute usually leaves open the details about how the law is applied and enforced. These details can be developed and issued as **regulations** through a process known as rule-making or through more informal guidance released by agencies. In short, Congress can use a statute to create administrative agencies and to give power to the agency to create **regulations** that are enforceable through law.

2. Administrative law. Administrative law is the body of law governing administrative agencies. Administrative agencies may issue **regulations** and enforce them. Agencies may conduct hearings regarding disputed matters relating to **regulations** or enforcement through a process called *adjudication*.

3. Executive orders. The Constitution vests executive power in the president of the United States. Part of this power allows the president to issue executive orders, and similar powers exist for state governors. Executive orders do have the force of law, but executive orders are limited in application to areas where the executive has authority.

4. Common law. Common law applies in the absence of constitutional or statutory law. It is judge-made law, and the outcome of a court case can set precedent for others that sue for similar claims.

In total, there are five sources of law: the Constitution, **statutes**, administrative law, executive orders, and common law. These five sources of law can be interpreted by judges, which is often referred to as *case law*. Case law is any evolution of law through court decisions.

11.3 Water in Context with US Law

Each of the previously mentioned branches and sources of law have played significant but different roles in shaping US water policy. Here we take a look at each of the sources of law and examples of their application to water quality and water quantity. The following examples are not extensive or fully inclusive of all the elements of law within each example. In law, nuances are the name of the game.

11.3.1 The Constitution and Water

Although the US Constitution does not have specific legal language linked to water quality or quantity, the Constitution may influence law of water quality and quantity. For example, the Constitution can play a role in water quality and water quantity disputes under the "**takings**" clause (Fifth Amendment). The **takings** clause requires that the government provide compensation when it seizes private property for public use, a practice referred to as **eminent domain. Takings** encompass actions that seize private property physically and nonphysically. That is, **takings** of private property can be either physical (e.g., a parcel of land taken by the

Box 11.3 The Edwards Aquifer

The Edwards aquifer in Texas in the United States provides water for irrigated agriculture, the city of San Antonio, and the discharge for large springs at Comal and San Marcos that are habitats for a diverse ecosystem, including several endangered species. In 1991, a suit was filed under the Endangered Species Act to limit unrestrained pumping of water from the aquifer for irrigation; the unrestrained pumping was decreasing spring flow, and thus the species depending on the flow were threatened. In 1993, the Texas legislature created the Edwards Aquifer Authority (EAA) to control pumping and protect the springs.

After applications for permits to expand historical pumping rates on the Bragg Pecan Farm were denied or restricted by the EAA, Bragg Pecan Farm filed a suit claiming a regulatory taking. After a protracted court battle, the courts ruled in favor of the plaintiff (the Braggs), and in February of 2016, a jury awarded $2.55 million in compensation for the regulatory taking.

A concern for the future is that any extension of this ruling to other cases could have a chilling effect on regulations to protect the environment.

government for public purposes) or regulatory (e.g., economic burdens resulting from government regulation).

Not surprisingly, both physical and regulatory **takings** are much-contested issues for water rights, especially in the Western states. Although the law on water rights and **takings** is still evolving, there are some instances where **takings** claims have been used successfully to obtain compensation for physical and regulatory **takings** (Box 11.3). Laws that prompt claims of **takings** typically are aimed at environmental protection, so there is a tension between public interests and those of individual property holders.

11.3.2 Statutory and Administrative Law and Water

Statutory law and administrative law are tied together closely. **Statutes** can be used to create administrative agencies and to authorize the agency to create **regulations** that are enforceable through law. These agencies have officials that conduct hearings and provide written decisions about disputed matters regarding such things as enforcement.

For a good example related to statutory law and administrative law in the context of water quality, consider the Clean Water Act. In 1948, the Federal Water Pollution Act, the first significant law to address water quality, was passed. This statute was amended in 1972 to incorporate more stringent criteria for controlling water pollution, and it subsequently became known as the Clean Water Act (CWA). The CWA has been amended many times since 1972, but its general purpose remains the same: to provide a framework for regulating water pollutants, maintain specified water quality standards for contaminants, and grant authoritative powers to the Environmental Protection Agency (EPA) to implement pollution control measures.

The CWA is statutory law, which gave the EPA, an administrative agency, the authority to write and enforce **regulations** regarding water pollutants. Before the CWA existed, water quality was, in the absence of other state legislation regulating water quality, largely regulated through common law. Judges have the ability to interpret elements of the CWA, such as the geographic extent or the scope of the CWA, establishing case law.

The EPA is a federal regulatory agency, with regional offices. The EPA oversees and implements much of the federal CWA. States can impose stricter laws than the CWA, but states must follow minimum standards for water quality management set out in the Clean Water Act. If they do not, the EPA may step in to remedy a state's shortcomings.

Although water quality is a matter of federal and state law, water quantity is mostly a matter of state law. States have used a combination of common law and statutory frameworks. In many states, the state engineer's office is the administrative agency of interest for matters of water quantity. For example, Nevada's statute NRS 533.325 indicates that water appropriations (exemptions apply) must be permitted by the State Engineer's Office.

> Application to State Engineer for permit. Any person who wishes to appropriate any of the public waters, or to change the place of diversion, manner of use or place of use of water already appropriated, shall, before performing any work in connection with such appropriation, change in place of diversion or change in manner or place of use, apply to the State Engineer for a permit to do so.

Statutes are useful in setting out a framework for water rights, but contention about water rights still can arise.

11.3.3 Executive Orders and Water

Executive orders are useful in promoting better management of water resources, especially with regard to issues that are time sensitive. The president of the United States and most state governors can issue executive orders where the executive has corresponding authority (Perkins, 2017). For example, in 2014, the governor of California ordered the first-ever statewide mandatory water reductions (Executive Order B-29-15(2), Executive Order B-29-15(10)).

> The State Water Resources Control Board (Water Board) shall impose restrictions to achieve a statewide 25% reduction in potable urban water usage through February 28, 2016. These restrictions will require water suppliers to California's cities and towns to reduce usage as compared to the amount used in 2013 . . . The Water Board shall require frequent reporting of water diversion and use by water right holders, conduct inspections to determine whether illegal diversions or wasteful and unreasonable use of water are occurring, and bring enforcement actions against illegal diverters and those engaging in the wasteful and unreasonable use of water.

This unprecedented action by the California governor granted additional authoritative powers to the state's regulatory agency so that the agency could impose restrictions to reduce water usage and increase enforcement against illegal uses.

Executive orders can be especially useful in setting new thresholds of water use within a state during periods of drought, when water is scarce and action is time sensitive. Nevertheless, authorization for executive orders varies significantly by state, and some states provide broad authority for executive orders to governors, while other states provide little to no authority for executive orders to governors (Perkins, 2017).

11.3.4 Common Law and Water

One of the challenges to achieving sustainable management of water resources is that water law, in many cases, evolved before the advent of current scientific knowledge about hydrology. The legal frameworks for stipulating the use of surface water and **groundwater** quantities are often distinct, even though there is scientific knowledge that the waters are interconnected. Furthermore, considerations of water quality typically are separate as well. The legal frameworks notwithstanding, surface water, **groundwater**, and water quality interact. Water law must evolve as conflicts among these domains arise (Box 11.4).

Box 11.4 Water Law and Connections between Surface Water and Groundwater Quantity and Quality

One example of legal conflict relates to the management of water resources in Siskiyou County, California. An issue that has been raised is how groundwater withdrawals relate to the "public trust," or the broad public interest in maintaining ecosystem health in contrast to the private right to use water. The Scott River in Siskiyou County is an important habitat for salmonid fish, including a few endangered species. These fish require not only that a minimum low flow be maintained but also that the water be kept at or below a maximum temperature.

In 2011, the Environmental Law Foundation and others filed suit contending that Siskiyou County had an obligation to regulate groundwater withdrawals to protect the public trust in terms of the Scott River. Prior to the suit, the public trust doctrine had not been applied to groundwater. Despite the arguments by Siskiyou County and the State Water Resources Control Board that the public trust doctrine did not apply to groundwater, the California Superior Court in 2014 ruled that groundwater had to be regulated in accordance with public trust principles. In 2018, the California Court of Appeals for the Third Appellate District upheld the applicability of the public trust doctrine to the Scott River. This ruling will be an important precedent that recognizes that pumping groundwater in connected groundwater–surface water systems has impacts on surface water quantity and quality (e.g., temperature) and on ecosystems. The recognition of hydrological principles within legal frameworks can allow shaping better laws and policies to manage water resources sustainably.

In many cases, water quality laws have transitioned from using a common law framework to a statutory and administrative framework that is largely built at the federal level. The reasons for this transition away from common law are numerous. Many critics of the common law approach did not think that common law adequately accounted for the public's newfound and strong interest in water quality during the 1960s and 1970s. Under common law, lawsuits were often filed after pollution had occurred. Without a regulatory body, courts had to rely on the plaintiff (see Box 11.2) to find pollution, and it can be difficult for plaintiffs to monitor water quality themselves. Because common law addresses issues on a case-by-case basis, the cumulative impacts on water quality were overlooked. Multipoint source pollution (Chapter 12) made it difficult to identify the cause or causes of the pollution, and both technical and scientific expertise remain critical for understanding the *who*, *what*, *where*, and *when* of contamination events. Many courts were, and are still, not adequately trained to resolve such discrepancies.

The federal government plays a more limited role in water quantity than water quality. As a broad generalization, water rights were developed through an initial era of no controls or very limited

Table 11.1 Surface water rights doctrines for water allocation simplified. Information based on Getches et al. 2015.

Doctrine	How is water allocated?	How is scarcity apportioned?
Prior appropriation	Priority, based on timing of use	Junior ("newer") water rights holders are first to reduce use during scarcity
Riparian	Sharing, based on land ownership among other considerations such as reasonable use	All users reduce water withdrawals with consideration to land ownership and other considerations such as reasonable use
Hybrid	Priority and sharing considerations	Based on land ownership and timing of use

controls on withdrawals or principles for dispute resolution used in practice. As use of water progressed, common law doctrines developed to deal with competition, and the arbiter of water rights was predominantly the court.

(A) Common Law Doctrines for Surface Water Allocation of surface water is governed by three approaches: **riparianism**, **prior appropriation**, and a hybrid of the two (Table 11.1) *Riparian rights* are allocated to water users adjacent to streams, rivers, or other water resources. The word "riparian" refers to anything related to the banks of a river, although **riparianism** is interpreted more broadly to include adjacency to other water resources. Basically, the riparian doctrine states that a person who owns property adjacent to a river has the right to reasonable use of water flowing past the location. Riparian rights evolved during an era of water abundance. Because rights are not prioritized under pure **riparianism** approaches, users share the burden of reduced allocations during times of scarcity. Two critiques of **riparianism** are that the doctrine does not account for the effects of consumptive use of water by upstream users on downstream rights and that the doctrine does not account for **groundwater**–surface water interdependencies. Furthermore, reasonable use limits on riparian rights can change over time, reducing certainty for water users and potentially limiting investment.

As distinct from **riparianism**, the doctrine of **prior appropriation** is based on the concept of *priority* of the use of water. That is, priority is a function of when water is first put to beneficial use or to when intent to put the water to use is established: first in time, first in right. The first party to put the water to beneficial use or show intent has senior rights, and the parties that follow have junior rights. Senior water rights receive their entire allocation before junior rights receive any water. Appropriative rights provide for a quantified amount of water. Because beneficial use provides the foundation of this system of rights, if the water right is not used, it can be forfeited (Goldfarb, 1991). This concept of "use it or lose it" can be a disincentive to improve efficiency of water use, but its original goal was to discourage speculation in water rights by owners who were not actively using their water rights. Unlike riparian rights, rights through the **prior appropriation** system are not based on land ownership. Anyone can obtain a right under the **prior appropriation** system, and in most states, water rights and land rights are unbundled. A key problem associated with the system is overappropriation—there are many cases where the entire flow of a river in dry years is appropriated or overappropriated (e.g., Grantham & Viers, 2014) for consumptive use, leaving little or nothing for environmental flows.

States that use hybrid systems fall into two categories: states that use both, like California, and states that originally used **riparianism** but eventually abandoned it for **prior appropriation**. Generally speaking, in the United States, the Eastern states use **riparianism**, the Western states use **prior appropriation**, and the Midwestern states use a hybrid of the two (Thompson et al., 2013).

The increasing recognition of a need for more intensive management and security has evolved as competition for water has increased. **Regulated riparianism** is a legislative approach—as opposed to a common law approach—that requires permits for some or all withdrawals. Permitting requires advanced permission to withdrawal water, and it is often designed to look at the water system as a whole as well as individuals within the system. In theory, if the permit system is set up, maintained, and enforced properly, it can be a proactive means of managing water withdrawals.

(B) Common Law Doctrines for Groundwater Five doctrines govern **groundwater** rights in the United States: **prior appropriation**, correlative rights, rule of capture, reasonable use, and restatement of torts (Bryner & Purcell, 2003; Thompson et al., 2013). Although most states only apply one doctrine, there are some states that apply multiple doctrines. For example, California uses **prior**

Table 11.2 Groundwater rights doctrines for water allocation simplified. Information based on Getches et al. 2015; Thompson, Leshy, and Abrams 2013.

Doctrine	Legal rights to allocation are based on	How is scarcity apportioned?
Prior appropriation	Timing of first use	Junior ("newer") water rights holders are first to reduce use during scarcity
Absolute ownership	Land ownership, unlimited ability to withdraw water under land owned	No liability for rights holders to reduce use unless extreme waste or malicious withdrawals occur
Reasonable use	Land ownership, ability to withdraw reasonable quantities for use on land overlying the aquifer	"Unreasonable" uses, with unreasonable broadly defined, may be restricted
Correlative rights	Land ownership, ability to withdraw water based on proportion of land overlying the aquifer	All users reduce use during scarcity
Restatement of torts	Land ownership, ability to withdraw reasonable quantities for use on or off land overlying the aquifer	No liability for rights holders to reduce use unless "unreasonable" uses cause harm or exceed owners reasonable share of water

appropriation and correlative rights. The differences among the five doctrines are highlighted during times of scarcity (Table 11.2).

1. **Prior appropriation** of **groundwater** follows the same approach described above for surface water. Priority is a function of when water is first put to beneficial use, so the first party to do so has a senior right, and the parties that follow have junior rights. During times of scarcity, junior rights are limited so that senior rights are fulfilled first.
2. *Correlative rights* provide for proportionally equal rights to landowners who overlay a common **groundwater** source so long as the water use is reasonable and beneficial to the landowner's property. Correlative rights are similar to riparian rights for surface water—sharing is the dominant rule. During times of scarcity, water restrictions are shared equitably among all users.
3. The *rule of capture*, or absolute ownership, provides landowners with an unlimited right to withdraw water from the **aquifer** under the landowner's property. The unlimited right can be irrespective of the impact on others unless there is an extreme circumstance, such as malicious pumping. The rule of capture, in its basic form, does not include limits for water use, even during times of scarcity. As a result, the rule of capture tends to be more immune from efforts to more sustainably manage water than other common law doctrines.

4. The *reasonable use* doctrine, also known as the "American rule," is a modified version of the rule of capture. The withdrawal of water must be for reasonable use on the overlying land, and the definition of reasonable use can be very broad.
5. The final category, referred to as *restatement of torts*, is a blend of approaches that uses several factors to determine appropriate water use. It is stated as a liability rule, but it can operate as a rights regime (Thompson, Leshy, & Abrams, 2013).

The five doctrines of **groundwater** law focus on individual rights to water. Under the common law, these rights are not quantified until there is an adjudication. An *adjudication* is a formal decision-making process where the court quantifies water rights—that is, through evaluation of a lawsuit. Sometimes administrative agencies quantify water rights before the courts issue a final order in an adjudication. Adjudication can look very different across the United States. In some cases, the formalization of water rights under the adjudication process can require significant time and money. This has been the case in California, where adjudication has been used in cases of **groundwater** disputes. **Groundwater** extractions from many California **aquifer** systems are unsustainable, and they have resulted in subsidence, seawater intrusion, **infrastructure** damage, and negative impacts to **groundwater**-dependent ecosystems. These negative impacts can cause disagreements about **groundwater** use among landowners and nonlandowners who are withdrawing water from the **aquifer**.

People in California are hopeful that the passing of the Sustainable Groundwater Management Act of 2014 will provide a framework for managing aquifers sustainably, and thus possibly reduce the occurrence of expensive lawsuits. Indeed, concerns about the regional sustainability of aquifers have shifted many states to adopt more proactive approaches than the common law. For example, states are adopting statutory approaches (i.e., **groundwater** management plans, withdrawal permitting) at the state and substate level with increasing frequency (e.g., Nelson & Perrone, 2016).

(C) Doctrine of Equitable Apportionment Disputes about water allocation can occur within states as well as between or among two

or more states that may or may not use the same legal doctrines for water rights. In some cases, states negotiate interstate compacts, which have the effect of federal law upon congressional approval, to address concerns and identify how water allocation will be considered. In other cases, the states may not come to an agreement, in which case one or more of the states can bring suit against the other(s) in the US Supreme Court. In lawsuits between or among states, the Supreme Court has original jurisdiction and runs as a trial court. The court can appoint a Special Master to evaluate evidence, prepare conclusions of law, and provide the court with a nonbinding recommendation (Getches et al., 2015). The Special Master need not have experience with water law or water resources.

Because of the vagueness surrounding water rights and the complexity of the science, the Supreme Court is often disinclined to hear disputes about water allocation (Getches et al., 2015). To date, the court has presided over only a handful of such cases. Until recently, all the cases focused on surface water or **groundwater**–surface water disputes between or among states. The first case to the Supreme Court was in 1907 and involved Kansas and Colorado—two states with different water rights regimes for surface water (Anderson et al., 1984). The court established the doctrine of equitable apportionment, or the "equitable division" of water among users (*Kansas v. Colorado*, 206 US 46 [1907]). Equitable division is not the same as equal amounts, and the interpretation of equitable apportionment has evolved throughout the past century as new cases have been heard.

Interstate disputes have historically focused on surface water or **groundwater**–surface water interactions, but **groundwater** has been the sole topic of a recent dispute between Mississippi and Tennessee. In Mississippi, water rights are determined by a permit system created by the state's 1985 Water Resources Act; in Tennessee, water rights are determined by common law doctrines that have been modified by **statutes** (Cain et al., 2017). Because the two states have different water rights regimes, equitable apportionment by the Supreme Court may be appropriate (Box 11.5). Nevertheless, the doctrine has never before been applied to **groundwater** disputes.

Box 11.5 Equitable Apportionment for Groundwater

In 2005, Mississippi filed suit in federal district court against Memphis Light, Gas and Water (MLGW) and the City of Memphis, claiming that MLGW has pumped and captured groundwater that otherwise would have been stored in Mississippi's portion of the shared aquifer. The district court held that it did not have jurisdiction to hear the case. The district court indicated that Tennessee must be a party to the suit because the allocation of water is within the state's jurisdiction. Thus, the Supreme Court has original and exclusive jurisdiction (*Hood ex rel. Mississippi v. City of Memphis*, 533 F.Supp.2d 646; N.D. Miss. 2008). Mississippi appealed, but the Fifth Circuit Court of Appeals upheld the lower court decision.

In 2010, Mississippi filed a lawsuit against Tennessee, Memphis, and MLGW. Because the lawsuit was now between two states, the case was presented to the US Supreme Court. Mississippi only sought equitable apportionment as alternative relief. The Supreme Court denied hearing the case without comment (*Mississippi v. City of Memphis*, 559 US 501 [2010]). The court did cite footnotes in two prior cases, which suggested that the court had denied hearing the case because equitable apportionment was appropriate and Mississippi needed to show that pumping by the defendant caused substantial harm.

In 2014, Mississippi filed another lawsuit against Tennessee, Memphis, and MLGW, which is before the Supreme Court (*Mississippi v. Tennessee*, No. 143 [US]). In this lawsuit, Mississippi is not seeking equitable apportionment. Mississippi claims to own all groundwater under its land. So although the aquifer is a shared resource, Mississippi claims that the water pumped by Tennessee originated within the state of Mississippi and the water would not have been available to Tennessee if MLGW pumping did not alter the natural groundwater discharge regime. Tennessee argues that the dispute is an equitable apportionment case because of the court's precedent.

Several aspects of the case have implications for water law and groundwater management. Historically, management of groundwater has lagged management of surface water. As a result, groundwater and surface water tend to be managed as separate resources, with separate water rights regimes. Because the *Mississippi v. Tennessee* case is the first time states are arguing over groundwater resources, as opposed to surface water or groundwater–surface water interactions, the Supreme Court's decision could inform whether interstate disputes over groundwater and surface water will be treated the same and whether equitable apportionment is the proper method of allocation of groundwater aquifers between or among states. The case could also inform whether a state can have an "ownership" right of the groundwater beneath its surface (Klein, 2018). The Supreme Court's ruling could have a significant influence on groundwater law and management moving forward.

11.4 A Global Look at Legal Frameworks

The legal approach to the management of water differs around the world. In this section we take a high-level look at how a handful of countries around the world approach water rights. In some countries water rights are formalized, but in others water rights are informally recognized with contemporary water rights. Customary water uses are the traditional water uses of a society, and these rights to water may be formally or informally recognized with contemporary water rights. Informal customary water rights are often supported by customary law or built upon oral agreements and local customs. As competition for water increases, priority can shift away from customary uses such as environmental and cultural flows or irrigation to water uses with higher economic return such as energy. This can impact customary water users whose rights may not be recognized within contemporary legal systems.

Some countries have recognized customary rights and have integrated these rights into their modern water-rights systems through formal mechanisms, while other countries have not (Ramazzotti, 2008). In Canada, indigenous community rights have been protected constitutionally since the early 1980s, and these rights can have priority over most other uses of water (Christensen & Lintner, 2006). In Sri Lanka and Thailand, customary water rights are not explicitly protected through legal frameworks, but in Sri Lanka customary rights are recognized in practice (Bird et al., 2009). In the United States, the reserved rights doctrine (Box 11.6) provides for written and formalized rights for Native American tribes. Nevertheless, tribal governments have sovereign status, and more work could be done to engage tribes on a government to government basis (Womble et al., 2018).

At the broadest and most generalized level, there tends to be a positive correlation between a country's economic status and their legal framework for water rights (Araral & Yu, 2013). That is, countries with developed economies tend to use formalized rules and standards to govern water use, and these formalized rights are more likely to be (1) enforced and (2) integrated with other environmental laws. One reason for this trend is the economic and resource burden surrounding the creation, maintenance, and enforcement of a legal system. Let us explore this generalization using two countries: the Philippines and Australia.

The Philippines was one of the earliest adopters of a permitting system in Asia. Since the passing of its 1976 Philippine Water Code, the country requires permits for both surface water and **groundwater**. Regardless of its well-regarded design, implementation of the permitting system has hindered its success. A study in the early 2000s—30 years after implementation of the permitting system—found that only 35% of required permit holders have a permit (Bird et al., 2009). The reasons cited for the poor adoption of permits include (1) lack of awareness, (2) poor coordination, and

Box 11.6 Reserved Rights in the United States

The *reserved rights doctrine* offers a unique opportunity to explore how the United States formalized some customary water rights. Reserved rights are federal water rights for Native American or public land reservations. These rights have priorities that date back to the establishment of a reservation. The doctrine of federal reserved water rights traces its origins to the decision of *Winters v. United States* (207 US 564 [1908]). In this case, the US Supreme Court ruled that when the United States sets aside a Native American reservation, it implicitly reserves sufficient water to fulfill the purposes of the reservation.

Over half a century later, the Supreme Court revisited reserved rights in *Arizona v. California* (373 US 546 [1963]). In this case, the court upheld reserved rights for Native American reservations. The ruling built upon the criteria for reserved rights by expanding reserved rights to all federally reserved public lands, such as national parks and forests. Subsequent court cases further defined such things as the amount of water reserved and the establishment of priority.

Reserved rights can be controversial, especially where water rights are overallocated. Rights to use water typically are allocated by states using common law or regulated by states using statutory law. Reserved rights, however, exist under federal law, which prevails over state law.

For example, states that use prior appropriation often establish priority based on beneficial use and "first in time, first in right." The priority of reserved rights, however, dates back to the establishment of the reservations, and these dates of establishment are most likely earlier than state priority dates. Because reserved rights are often not quantified when the reservation is established, federal claims to reserved rights that are acknowledged retroactively can unsettle state allocation systems. Reserved rights do not fall under the power of eminent domain, so there is no compensation for rights that are taken away for reserved right purposes.

Before 2017, it was not clear whether tribal groundwater rights were considered reserved rights. But, in late 2017, the US Supreme Court left standing a ruling that endorsed tribal groundwater rights as reserved rights (*Agua Caliente Band of Cahuilla Indians v. Coachella Valley Water District*, 2017). This ruling has the opportunity to promote tribal water security in the United States, as well as sustainable groundwater management (Womble et al., 2018).

(3) resource limitations of the enforcement agency—all of which underscore the importance of financial investments and leadership to create, maintain, and enforce legal systems.

Australia is a prime example of a water rights framework that has been designed and implemented successfully (although it still faces challenges). In Australia, early water rights were based on riparian rights. The system of water rights evolved quickly, abolishing **riparianism** in favor of a statutory approach that vested water in the Australian states. In response to interstate water disputes and growing concerns about environmental and economic sustainability, water reform began to transition to a nationally coordinated effort in the late 1900s and early 2000s. In the present day, Australia's state **statutes** define water rights using a transparent system of allocation. State and federal statutory management plans affect how water rights can be exercised. Most rights are treated separately (e.g., unbundled) from land and are tradable, and many are enforced through metering, measuring, and monitoring.

11.5 Concluding Remarks

Legal systems for water are complex and not the sole opportunity available for sustainable **water management**. In some places and at certain times, water policies using legal frameworks may work, while in other places or at other times water policies focused on collective action or economic pricing may be more successful. There is no single remedy to tackle water scarcity. Managing water resources in a sustainable manner requires a diverse portfolio of tools customized to the needs of each place, yet adaptable to changes over time.

11.6 Key Points

- Water policy in the United States is complex and involves aspects of both water quality and water quantity. (Section 11.2)
- There are five sources of law in the United States (Section 11.2), all of which play a role in water law. (Section 11.3)
- The legal frameworks used to manage water quantity and quality vary across the globe. (Section 11.4)
- The creation, maintenance, and enforcement of a successful water law system requires economic and resource investments. (Section 11.4)

11.7 Example Problems

Problem 11.1. How does the regulation of water quality differ from the regulation of water quantity?

Problem 11.2. What are reserved rights? What is their priority?

Problem 11.3. In many states, we manage surface water differently than groundwater. In what hydrologic circumstances could this be problematic?

Problem 11.4. What is the most significant difference between the rule of capture and the other doctrines that govern **groundwater** rights?

Problem 11.5. Identify three reasons water quality laws have transitioned from common law to an administrative framework.

Problem 11.6. What is one reason for the positive correlation between a country's economic status and its legal framework for water rights?

Problem 11.7. Contaminants that affect the quality of our water resources can be geogenic (i.e., naturally occurring), **anthropogenic** (i.e., human caused), or both. Identify a contaminant of each type—geogenic, **anthropogenic**, and both—and identify their respective EPA standards covered under the CWA.

Problem 11.8. Identify the doctrine(s) that govern(s) surface water and **groundwater** rights in your home state.

11.8 Suggested Readings

Grantham, T., & Viers, J. (2014). 100 Years of California's water rights system: Patterns, trends and uncertainty. *Environmental Research Letters, 9*(8): 84012.

Womble, P., Perrone, D., Jasechko, S., Nelson, R. L., Szeptycki, L. F., Anderson, R. T., & Gorelick, S. M. (2018). Indigenous communities, groundwater opportunities. *Science, 361*(6401): 453 LP-455.

11.9 References

Anderson, S. T., Harder, T. A., Laskaris, N. Y., & Mittag, L. M. (1984). Equitable apportionment and the Supreme Court: What's so equitable about apportionment? Symposium on Minnesota Water Law: Research project. *Hamline Law Review, 7*, 405–429.

Araral, E., & Yu, D. J. (2013). Comparative water law, policies, and administration in Asia: Evidence from 17 countries. *Water Resources Research, 49*(9), 5307–5316.

Bird, J., Arriens, W. L., & Von Custodio, D. (2009). *Water rights and water allocation: Issues and challenges for Asia*. Manila, Philippines: Asian Development Bank. https://www.adb.org/sites/default/files/publication/28482/water-rights.pdf

Bryner, G. C., & Purcell, E. (2003). *Groundwater law sourcebook of the western United States*. Denver: Natural Resources Law Center, University of Colorado School of Law.

Cain, R. L., Goll, M., Hood, T., Lauer, C., McDonough, M., et al. (2017). *Groundwater laws and regulations: A preliminary survey of thirteen US states*. G. Eckstein & A. Hardberger (Eds.). Fort Worth: Texas A&M University Law School. https://law.tamu.edu/docs/default-source/faculty-documents/groundwater-laws-reg-13states.pdf

Christensen, R., & Lintner, A. (2006). Trading our common heritage? The debate over water rights transfers in Canada. In K. Bakker (Ed.), *Eau Canada: The future of Canada's water*. Vancouver: UBC Press.

Getches, D. H., Zellmer, S. B., & Amos, A. L. (2015). *Water law in a nutshell* (5th ed.). St. Paul, MN: West Academic.

Goldfarb, W. (1991). *Water law* (2nd ed.). Chelsea, MI: Lewis.

Grantham, T., & Viers, J. (2014). 100 years of California's water rights system: Patterns, trends and uncertainty. *Environmental Research Letters*, *9*(8), 84012.

Klein, C. A. (2018). Owning groundwater: The example of *Mississippi v. Tennessee*. *Virginia Environmental Law Journal*, *35*(474), 1–40.

Perkins, H. (2017). Chapter 4: State executive branch (Table 4.5). In *The book of the states 2018*. Washington, DC: Council of State Governments. http://knowledgecenter.csg.org/kc/content/book-states-2018-chapter-4-state-executive-branch

Ramazzotti, M. (2008). *Customary water rights and contemporary water legislation: Mapping out the interface*. FAO Legal Papers Online 76. Food and Agriculture Organization of the United Nations. http://www.fao.org/3/a-bb109e.pdf

Thompson, B., Leshy, J., & Abrams, R. (2013). *Legal control of water resources: Cases and materials* (5th ed.). St. Paul, MN: West Academic.

Womble, P., Perrone, D., Jasechko, S., Nelson, R. L., Szeptycki, L. F., et al. (2018). Indigenous communities, groundwater opportunities. *Science*, *361*(6401), 453–455.

CHAPTER 12

Water Quality

12.1 Introduction

Material in previous chapters described the **hydrological cycle** and water resources requirements for agriculture, energy production, households, and healthy ecosystems. The focus has been primarily on the amount of water available and required—that is, on measures of water quantity. There is also a relationship between the use of water and its impact on the *quality* of water (Table 12.1), and the use of water can be impacted by the physical and chemical state of water (Table 12.2). Broadly speaking, there are two categories of contaminants: anthropogenic and geogenic. Anthropogenic contaminants are caused by human activities. *Geogenic contaminants* occur naturally and are not caused by human activities. For example, some deep **groundwater** is naturally saline, and this saline **groundwater** may not be of use to domestic or agricultural users. Although geogenic contaminants are not caused by human activities, geogenic contaminants can be mobilized by human activities. For example, arsenic is found naturally in some subsurface geological materials, and extensive pumping can mobilize arsenic, resulting in **groundwater** that is contaminated with arsenic.

In this chapter we focus on some **anthropogenic** *contamination* mechanisms to our water resources as well as the impact of **anthropogenic** contamination on water users. **Anthropogenic** water contamination mechanisms are wide ranging; contaminants are introduced into streams, lakes, and **groundwater** through anticipated discharges and through accidental spills. Discharges that occur at one place, such as the outfall from a sewage treatment plant or a **thermoelectric** power plant, are termed **point discharges**. Discharges that occur across broad swaths of the land, such as from the runoff from agricultural fields, are termed **non-point discharges**. **Point discharges** are more easily identified

Table 12.1 Examples of the use of water and its impact on the quality of water.

	Selected impacts on water quality	
Use category	Direct	Indirect
Agriculture	Runoff of sediments to streams Discharge of nutrients, herbicides, and pesticides to surface and groundwater	Changes in land cover leading to increased erosion Clearing of riparian vegetation leading to increases in stream temperature
Electricity production	Discharge of boiler and cooling process wastes Coal pile runoff Thermal pollution Changes in temperature and oxygen content of water due to operation of hydropower facilities	Changes in land cover leading to increased erosion Discharge of contaminants from extraction of fossil fuels, including oil field brines and acid drainage from coal mines Discharge of refinery wastes, including refining of biofuels
Domestic water use	Discharge from residential sources including toilets, sinks, bathing, and laundry	Leaching of sludge from wastewater treatment plants

Table 12.2 Water quality impacts on water use.

Use category	Selected water quality impacts on use
Agriculture (irrigation)	Salinity—salts reduce water availability to plants Elements such as sodium reduce water infiltration rates Some toxic elements such as cadmium concentrate in plants making them unsuitable for consumption
Electricity generation	Dissolved solids in water lead to corrosion or scale formation Organic matter can lead to growth of harmful microbes
Domestic water use	Untreated water harboring disease-causing microorganisms or toxic substances Contaminants such as pharmaceuticals not removed in water treatment plants

than **nonpoint discharges**, and their impact on water quality may be more easily monitored than the impact on water quality from **nonpoint discharges**.

The types of contaminants introduced into natural water bodies are varied. Among the most common pollutants from the water-use sectors we have examined are nutrients from applied fertilizers, pesticides, herbicides, **thermal pollution** from warm or hot water, dissolved inorganic constituents, organic wastes, and pathogenic microorganisms. Nutrients, pesticides, and herbicides primarily are associated with agriculture but also with municipal water use; significant amounts of nutrients, pesticides, and herbicides occur in runoff from urban areas where lawns and shrubs are maintained with the help of chemical additives. **Thermal pollution** is associated primarily with electricity generation

(Chapter 6); **thermal pollution** is the discharge of water to a river system that is notably warmer than its natural conditions. Inorganic pollutants, such as heavy metals, result from cleaning of equipment in electricity generation and also from runoff from roads and highways. Organic wastes and pathogenic bacteria can be found in the discharge from sewage treatment plants, from agricultural operations including animal feedlots, and from water used in electricity generation operations where organic cleaning compounds are used.

Just as the types of contaminants are varied, the impact of **anthropogenic** contamination on water users varies. Drinking water quality requirements must consider nutrients, dissolved inorganic constituents, organic wastes, and pathogens, but these pollutants are less impactful on other users, such as thermoelectric power plants. Thermoelectric power production can be impacted by warmer river water, because plant efficiency is reduced when intake water for cooling processes is too warm. Contaminated water for agricultural use can directly impact crops (e.g., nutrients) or impact products for human consumption (e.g., pathogenic bacteria). Water quality for the environment must consider all types of pollutants, but some pollutants may be more relevant to certain plant and animal species than others. **Thermal pollution**, for example, can significantly and directly impact aquatic ecosystems but is less hazardous to terrestrial animal populations than aquatic ecosystems.

12.2 Impact on Water Quality from Water Users

There are thousands of chemical compounds that are used for various purposes and that subsequently find their way into surface water and **groundwater**. New chemicals of potential concern are being introduced into the market all the time. It would be impossible to discuss the current state of water resources with regard to all contaminants. We will concentrate on three major categories of contamination that arise in conjunction with the water use domains that we have discussed: (1) discharge of organic waste with main contributions from treated or untreated sewage linked to, for example, domestic water use and animal feedlots; (2) discharge of nutrients from fertilizers applied to agricultural fields and to lawns and open spaces in urban and periurban areas; and

(3) discharge of warm water to rivers from thermoelectric power plants that use **once-through cooling** systems.

12.2.1 Organic Waste

One of the main impacts from the discharge of organic waste to water bodies is the depletion of oxygen in the water. Organic material released into natural waters serves as an energy source for microorganisms (e.g., bacteria). The microorganisms grow rapidly given a "food" source and consume oxygen in their respiratory process. The amount of oxygen required to metabolize the organic waste is measured as the **biochemical oxygen demand (BOD)**. Discharge of large quantities of waste with high **BOD** leads to depletion of oxygen in the water. The resulting *anoxic conditions* lead to fish kills, emission of foul smells associated with gases such as hydrogen sulfide, the loss of recreational use of the waterway, and the diminishment of the aesthetic value of the waterscape.

BOD-containing waste can be consumed effectively by natural processes without noticeable environmental consequences when the amounts are within the **assimilative capacity** of the receiving water. Humans have always disposed of wastes into waters—the notion that "dilution is the solution to pollution" has been, implicitly if not explicitly, a primary concept of waste disposal for a long time. When the amount of waste disposed at any one time at any one place is relatively small, the **assimilative capacity** of the receiving waters is sufficient to degrade the organic waste. As the population of a city grows, however, the capacity of receiving waters is easily exceeded, leading to undesirable outcomes. For example, an extreme case is the infamous Great Stink caused by sewage in the Thames River in London in the summer of 1858. The Great Stink led Parliament to flee the city and, later, to commission work to clean the river (Halliday, 2001).

In developed countries, wastewater treatment was mandated in the twentieth century, leading to improvements in river water quality by reducing pollution by sewage. In the United States, the levels of **BOD** in rivers decreased steadily after passage of the Clean Water Act in 1972 (Keiser & Shapiro, 2018). In Europe, improvements in wastewater treatment have led to overall declines of **BOD** in rivers (Figure 12.1). Although there are regions in the developed world where **BOD** loads to surface waters are high, the widespread

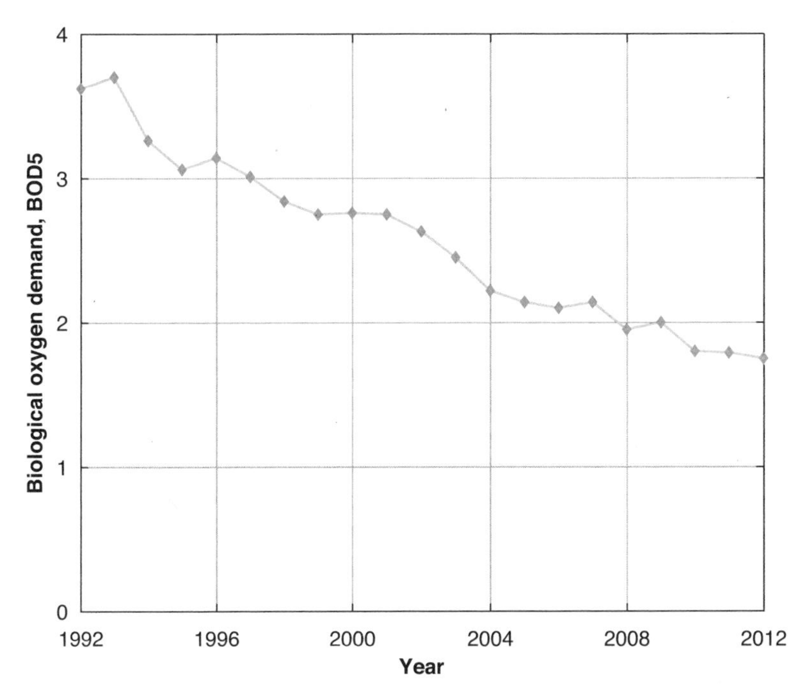

Figure 12.1 Biochemical oxygen demand (BOD) in European rivers, which has declined steadily over the past several decades. BOD5 is a measured oxygen demand made over a 5-day period. Data from European Environmental Agency 2012.

use of municipal wastewater treatment is effective in limiting **BOD**. Therefore, intensive livestock farming is the main remaining source of **BOD** discharges.

In parts of the developing world, however, wastewater treatment is insufficient, and many river segments have high **BOD** concentrations (Wen et al., 2017). Furthermore, as populations in cities in Asia and Africa grow and intensive livestock farming also increases, it is likely that the increased **BOD** load in rivers will lead to an increased risk of pollution from organic wastes for billions of people (Veolia and IFPRI, 2015; Wen et al., 2017).

12.2.2 Nutrients

The discharge of large quantities of nutrients to **groundwater** and rivers increased dramatically in the latter half of the twentieth century as fertilizer use in the industrial agricultural sector became widespread in many countries. Fertilizers are applied to agricultural fields, to lawns, to golf courses, and to other landscapes to stimulate the growth of crops and turf grass. Fertilizers such as ammonium and potassium nitrate, used to supply nitrogen to plants, are dissolved in water. A portion of the chemicals are transported to streams by surface runoff or percolation through the

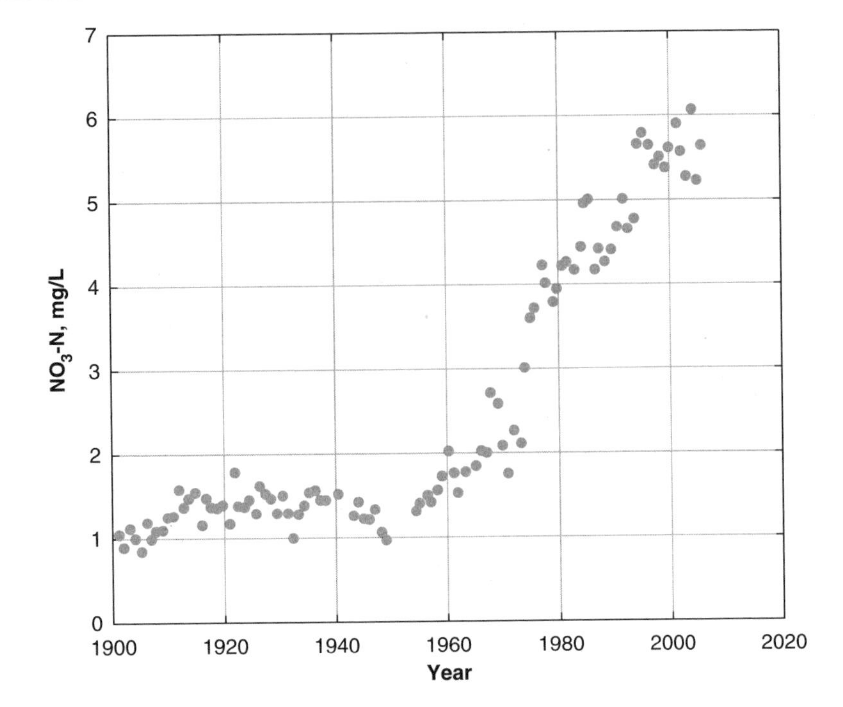

Figure 12.2 Nitrate concentrations in the upper Seine River, France. Data from Meybeck et al. 2018.

unsaturated zone to the **water table** and thence by **groundwater** discharge to streams or other surface waters. The fertilizers that stimulate the growth of land plants also stimulate the growth of algae and other aquatic plants once they reach streams, rivers, and lakes. **Eutrophication**, the overenrichment of natural waters with nutrients, can lead to unsightly blooms of algae that clog waterways. Often the harmful algal blooms include algae (cyanobacteria) that produce toxins. When the algae die, they become an internal **BOD**. This can result in what is known as *dead zones*, when the extensive depletion of oxygen makes the water unfit for use and can result in fish kills.

Nitrogen compounds are an example of unrestrained use leading to serious pollution issues. In the Seine River Basin in France, for example, nitrate concentrations increased steadily over a 50-year period. In 1950, there was a little more than 1 mg/L nitrate as nitrogen (NO_3-N), but by 2000 the nitrate concentrations peaked at about 6 mg/L NO_3-N (Figure 12.2). The nitrate pollution was stabilized at the peak level only with aggressive control measures (Meybeck et al., 2018).

In the United States, nitrate concentrations in rivers tended to stop increasing in the last decade of the twentieth century as the intensity of agricultural use of land declined in some areas and as the inputs from livestock and fertilizer decreased somewhat (Stets et al., 2015). Nevertheless, significant improvements in water quality may not be realized for decades. Because nitrate is soluble and does not adsorb readily to soil particles, it is transported to **groundwater** from where it continues to flow into streams for decades or longer. Even when control measures are taken to stem the use of nitrogen fertilizers, it can take a long time to see improvements in surface waters. For example, the nutrient runoff from the Mississippi River to the Gulf of Mexico that causes a large dead zone will persist for decades even if fertilizer application in the large agricultural areas in the Mississippi River Basin are curtailed. This is due to the legacy of previous decades of fertilizer application (Van Meter et al., 2018).

The flux of nitrogen in rivers has increased worldwide over the past several decades, with major inputs from regions in India and the People's Republic of China (Seitzinger et al., 2010). This reflects the observed rapid increases in population and agricultural intensity in India and China. Population and agricultural intensification are expected to continue increasing in many countries in the future. Globally, changes in climate (e.g., intense rainfall) are also anticipated. Consequently, forecasts are for worsening conditions of nutrient runoff with the attendant problems they cause (Seitzinger et al., 2010; Sinha et al., 2017) unless aggressive nutrient management options are adopted.

12.2.3 Thermal Pollution

As discussed in Chapter 6, **thermoelectric** power plants that use **once-through cooling** often withdraw significant amounts of water from rivers. Most of this intake water is returned to the river. Because the purpose of the intake water (i.e., cooling water) is to dissipate heat from the power plant, the outflow to the river is returned at a higher temperature than the intake from the river. Recent studies suggest that typical cooling-water outflows from once-through facilities are approximately 10°C warmer than the intake water (Madden et al., 2013). The impacts of **thermal pollution** are most acute during warm months, immediately near the **discharge** points, and where and when river flow rates are low.

Thermal pollution from power plants affects rivers around the world, and the United States, Europe, and China are especially

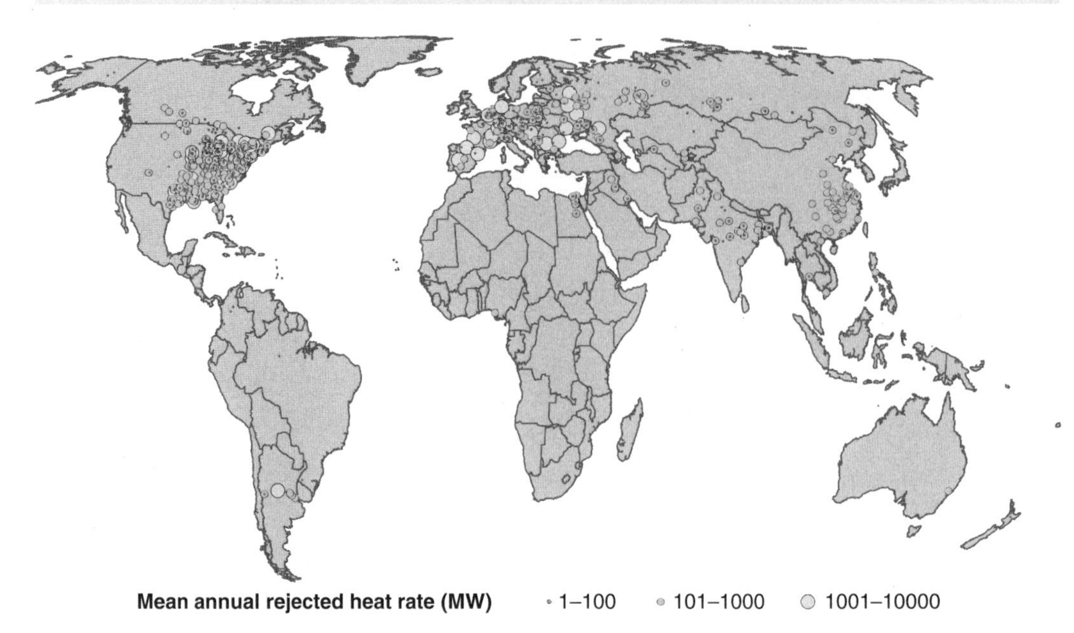

Figure 12.3 Thermoelectric power plants with once-through cooling reported in 2012. Thermal pollution can be inferred by the amount of heat rejected from the power plant, which is scaled by the dot size and color. Data from Raptis and Pfister 2016.

impacted. These areas have the lion's share of global power plants using **once-through cooling** (Figure 12.3). In the United States and Europe, regulatory limits have been placed on the discharge of heated water. One consequence of the imposition of regulations is phasing out of **once-through cooling** in favor of **recirculating** or **dry cooling** (Raptis et al., 2017).

12.3 Impact on Water Users from Water Quality

Managing water resources with an eye toward minimizing water pollution while meeting the demands of the global population is a goal that must be met to avoid potentially catastrophic consequences for ecosystems and humans. Many of the activities that reduce water quality also can be impacted by poor water quality. These impacts are wide ranging and can have varying consequences on the spectrum of water users.

12.3.1 Organic Waste

High **BOD** can restrict the use of water for municipal supply. There are no absolute standards for the level of **BOD** allowable in drinking water, but high values—above a few milligrams per liter—

are considered a possible indicator of contamination by pathogens associated with human and animal waste. **BOD** indicates presence of organic carbon, which also can be detrimental when drinking water supplies are treated using chlorine to disinfect the water. Chlorine combines with organic carbon to form dissolved chlorinated compounds (e.g., trihalomethanes), which are known carcinogens.

12.3.2 Nutrients

The impact of nitrate contamination on the use of water can be significant. High nitrate (NO_3-N) levels can also cause damage to **ecosystem services** through creation of zones of *hypoxia*, or areas with very low dissolved oxygen. In the United States, the EPA sets a drinking water limit of 10 mg/L NO_3-N to prevent "blue baby" syndrome. Recent studies suggest that levels as low as about 3.9 mg/L NO_3-N are associated with an increased risk of cancer (Schullehner et al., 2018).

In rivers in Iowa, nitrate concentrations increased in the second half of the twentieth century. Currently, nitrate concentrations typically are in the range of 6 mg/L NO_3-N (Kelly et al., 2015). It is little wonder that in 2015 the city of Des Moines sued upstream counties in Iowa to reduce fertilizer runoff into the city's drinking water supply. The suit was dismissed in the courts, and the city now expects to invest millions of dollars to upgrade its water treatment facilities to remove nitrate from its drinking water sources.

12.3.3 Thermal Pollution

Increased water temperatures can be especially dangerous to aquatic ecosystems. Warm temperatures can stress aquatic species by reducing reproduction rates and biological processes, and increasing their susceptibility to contaminants (Heugens et al., 2001). In some cases, **thermal pollution** can increase river temperatures so significantly that the water reaches a critical maximum temperature for some species, resulting in death.

Thermal pollution may also have an impact on power plants downstream, especially in areas where many power plants line a river within a short distance. Higher intake water temperatures can reduce the efficiency of a **thermoelectric** power plant. In countries with river water temperature regulations, high temperatures may require power plants to curtail power generation or shut down. Recent work focused on the Mississippi River watershed has indicated the need for a systemwide approach to evaluate the impacts on power production due to **thermal pollution** from upstream **thermoelectric** power plants (Miara et al., 2018) and the trade-offs associated with conversion to alternate cooling technologies.

12.4 Challenges into the Future

Over the past century, society has made advances in our technology to treat poor quality water so that it is usable. In some places, legal tools have been developed to protect water from **anthropogenic** contaminants (Chapter 11). Nevertheless, there are still water quality challenges that limit access to safe drinking water, impact ecosystems, and reduce the efficiency of our industries.

12.4.1 Organic Waste

Domestic, agricultural, and industrial activities contribute organic wastes to water bodies. Each of these will intensify in the future as the global population grows. The population has grown dramatically and will continue to do so for the next several decades (Chapter 9). Changing climate (Chapter 10) is likely to exacerbate the water quality impact of waste discharges. Without greater attention to pollution prevention, the **BOD** loading to freshwater systems is projected to increase by 9% to 32% (Veolia & IFPRI, 2015). Limiting the discharges of inadequately treated human and animal wastes in the future is a challenge that must be met to protect ecosystems and human health.

In developed countries, municipal wastewater streams are sent through sewers to a treatment plant where solids are removed (primary treatment) and most of the remaining organic pollutants are oxidized (secondary treatment). In developing countries, wastewater streams are often disposed untreated into rivers or the ocean. Half of the world's population now lives in cities. This fraction is expected to grow to 60% by 2040. While urbanization can offer improved access to services for safe drinking water and sanitation, the development of **infrastructure** such as treatment facilities has yet to keep pace in parts of southeast Asia and in Africa (Box 12.1). It will be necessary both to invest in **infrastructure** and to develop

Box 12.1 Point-of-Use Treatment

A key component of a municipal water system is raw water treatment (see Figure 7.1). This often requires high capital investment in infrastructure, such as a water treatment facility that filters suspended particles and disinfects the water so it is safe for drinking (see Box 7.1). In areas with limited access to financial resources or in areas that are rural, centralized treatment facilities may not be feasible. In these areas, *point-of-use treatment* may be applied. Point-of-use treatment is a decentralized approach to treating water, where the end user treats the water before using it. In some cases, point-of-use treatment may be more efficient than a centralized treatment facility, because not all water uses require water to be of potable quality. Nevertheless, point-of-use treatment places the burden on the end user, which may be especially burdensome for disadvantaged households.

Point-of-use treatment includes a variety of options, such as boiling, chemical disinfectants, and solar disinfection (Mintz et al., 2001). Boiling water is effective, but it may not be sustainable for communities that have limited fuel. Chemical disinfectants, such as sodium hypochlorite, can be effective at eliminating bacteria but may not eliminate parasites and viruses like boiling water can. Sodium hypochlorite is low cost and has a residual effect, meaning that water is protected for some time after treatment. Some chemical disinfectants can affect the taste and odor of treated water, potentially affecting perceptions of the water. Solar disinfection harnesses solar radiation to treat water. In Kenya, clear plastic containers that can transmit ultraviolet particles are used to treat small to moderate volumes of water (Mintz et al., 2001). Compared with chemical disinfectants, solar disinfection is less expensive, more accessible, and does not alter the odor or smell of the water.

social and administrative programs to sustain initiatives that ensure continuous adequate treatment.

Livestock production over the past several decades has grown rapidly in response to population growth and to the increased economic well-being of people in many countries. This growth is projected to continue. For example, meat consumption is one indicator of the past and projected future demands on livestock production (Figure 12.4). Given broad experience where livestock wastes have caused serious pollution of surface waters and **groundwater** in the past, it is imperative that liquid and solid wastes from livestock be managed in the future to meet sustainability goals. Both technological (e.g., advanced manure treatment) and social-cultural (e.g., changes in dietary preferences) solutions will be needed to address the issue.

12.4.2 Nutrients

The same trends and drivers that affect organic wastes also affect nutrients. Nutrients are applied to croplands to improve yields. Expansion of food production will be essential to feed the world's

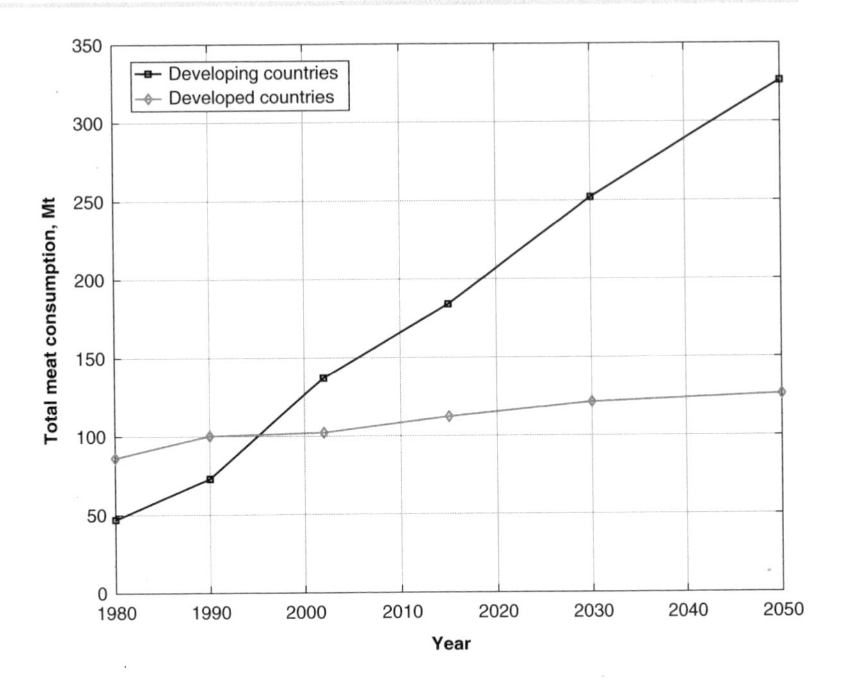

Figure 12.4 Meat consumption trends in million tonnes (Mt). Data from Thornton 2010.

population in the future. The Green Revolution staved off the Malthusian catastrophe (Chapter 9) but did so at a cost of increased production of synthetic fertilizers through, for example, a process to fix nitrogen from the atmosphere into usable fertilizer. Over the past five decades the loss of nitrogen from agriculture increased by 500% (Lassaletta et al., 2016). Without intervention to curb nitrogen pollution, projections indicate that nitrogen pollution of fresh waters will increase by 35% to 62% by 2050 (Veolia and IFPRI, 2015).

Nitrogen contamination of water supplies already is serious, and additional loads on the system will place large populations at risk from pollution. Avoiding the worst of the consequences while still providing adequate food for a burgeoning population presents a complex challenge. Once again solutions will involve a mixture of technological and social-cultural adaptations.

12.4.3 Thermal Pollution

Changes in precipitation and increased temperatures due to climate change are likely to increase river temperatures and reduce river flow rates during the summer season. The summer tends to coincide with peak electricity demand, suggesting that changes in climate could exacerbate the impact of **thermal pollution** on ecosystems and **thermoelectric** power plant efficiency. This challenge may be more significant in places that are already facing **water scarcity**. For example, northern China, a water-scarce region, is also dotted with once-through power plants (see Figure 12.3). Even places that on average have large amounts of rainfall may face **water scarcity** challenges as demands for surface water increase and regulations to protect ecosystems are realized.

A move to recirculating or dry cooling or **renewable energy** sources that do not require cooling water is likely to be a key adaptation strategy. For older power plants that need to be retrofitted, the costs may be considerable. A recent study looking at retrofitting all coal plants in Texas with **dry cooling** systems found it could save about 5.5 billion m^3 of water annually, but at a cost of about \$1.80 per m^3 of water saved (Loew et al., 2016). A cost of about \$1.80 per m^3 of water saved is about the same as the cost of highly treated residential water supplied to consumers in the United States.

12.5 Concluding Remarks

Many problems associated with water pollution already have clear solutions. Cropping decisions, no-till agriculture, and reducing the application of synthetic fertilizers have beneficial effects that include the reduction of pollution by nutrients. We know how to provide sanitation, which is essential to avoid major water pollution outcomes from discharges in urban centers. An integrated solution to the sanitation problem will build on both existing and new technologies and will involve recycling, including recovery of energy and nutrients from wastes (Andersson et al., 2016). In the sphere of agriculture, technological advances, such as coating fertilizer with polymers to enhance nitrogen use efficiency by crops (Snyder, 2017), provide opportunities for managing for improved water quality.

Employing technologies to address the challenges of meeting water quality goals require significant investment in **infrastructure**. Gaining the political will for these investments is often a challenge, particularly in rapidly growing, developing countries. The main constraint in resolving water quality problems in the future is likely to be the same as for other water resources management issues—the human component of **water management**, not the technical one (Cosgrove & Loucks, 2015). Developing adequate governance procedures and policies globally to manage our water quality problems will be equally important, or even more important, than providing the technologies to meet the challenges.

12.6 Key Points

- Discharges that occur at one place, such as the outfall from a sewage treatment plant or a **thermoelectric** power plant, are termed **point discharges.** (Section 12.1)
- Discharges that occur across broad swaths of the land, such as from the runoff from agricultural fields, are termed **nonpoint discharges.** (Section 12.1)
- Water quality often determines appropriate uses of water. For example, drinking water quality requirements must consider nutrients, dissolved inorganic constituents, organic wastes, and pathogens. (Section 12.1)
- The amount of oxygen required to metabolize an organic waste is measured as the **biochemical oxygen demand (BOD).** (Section 12.2.1)
- **BOD**-containing waste can be consumed effectively by natural processes without noticeable environmental consequence when the amounts are within the **assimilative capacity** of the receiving water. (Section 12.2.1)
- The discharge of large quantities of nutrients to groundwaters and rivers increased dramatically in the latter half of the twentieth century as fertilizer use in industrial agricultural became widespread. The large quantities of nutrients have led to **eutrophication**, the overenrichment of natural waters with nutrients, in many places. (Section 12.2.2)
- **Thermal pollution** is the discharge of water to a river system that is notably warmer than its natural conditions. Typical cooling water outflows from once-through thermal electricity generating plants are approximately 10°C warmer than the intake water. (Section 12.2.3)
- High levels of **BOD**, high levels of nitrate, or elevated water temperatures from waste discharges can negatively affect aquatic ecosystems and can restrict the use of water for municipal supply. (Section 12.3)
- Limiting the discharges of inadequately treated human and animal wastes in the future is a challenge that must be met to protect ecosystems and human health. (Section 12.4.1)

• Minimizing nutrient pollution of fresh waters while still providing adequate food for a burgeoning population presents a complex challenge for water managers. Solutions to the problem will require a mixture of technological and social-cultural adaptations. (Section 12.4.2)

• Transitioning from once-through to recirculating cooling or dry cooling is likely to be a key strategy for resolving **thermal pollution** problems, but this strategy is likely to increase costs. (Section 12.4.3)

• Developing adequate governance procedures and policies globally to manage our water quality problems will be equally important as infrastructural and technological solutions. (Section 12.5)

12.7 Example Problems

Problem 12.1. Wastewater treatment in developing countries has not been keeping pace with population growth. One basic and relatively low-cost method of treatment is a waste stabilization pond, which essentially holds wastewater to allow natural degradation processes to remove a large fraction of the **BOD** and a fraction of nutrients. In Nairobi, delivery of drinking water to inhabitants is a huge problem (Chapter 7). The collection and treatment of wastewaters is similarly deficient. The Dandora facility can process 80,000 m^3 of wastewater per day. Currently, Nairobi's population is 3.8 million people, 80% of whom live within the access area of the Dandora waste stabilization pond.

a. The average per capita waste generation is 125 liters per day. What percentage of Nairobi's population within the access area can be served by the Dandora facility?

b. How many additional facilities the size of Dandora would be required to meet the complete sanitation needs of the current population within the service area?

Problem 12.2. Consider the investment requirement in **infrastructure** for a waste stabilization pond system to be $30 per person served. (Note that this cost estimate is for the treatment facility itself and not for sewerage, such as the pipes. The total cost might be double the estimate for the facility costs.)

a. What would be the cost to provide basic treatment for the currently unserved population of Nairobi (see the previous question for details) using waste stabilization pond technology? Consider only the investment cost for the treatment facilities.

b. The population of Nairobi is projected to grow from 3.8 million in 2019 to 8 million by 2035 (Chapter 7). Assuming that the current population could be served fully, what would be the investment needed for additional waste treatment facilities each year to keep pace with the population growth? (Consider only the base facility costs and not inflation or interest rates.)

12.8 Suggested Reading

Veolia and IFPRI. (2015). *The murky future of global water quality.* Washington, DC: International Food Policy Research Institute. http://www.veolianorthamerica.com/sites/g/files/dvc596/f/assets/documents/2015/04/IFPRI_Veolia_H2OQual_WP.pdf

12.9 References

Andersson, K., Dickin, S., & Rosemarin, A. (2016). Towards “sustainable” sanitation: Challenges and opportunities in urban areas. *Sustainability, 8*(12), 1289.

Cosgrove, W. J., & Loucks, D. P. (2015). Water management: Current and future challenges and research directions. *Water Resources Research, 51*(6), 4823–4839.

European Environmental Agency. (2012). *European waters: Assessment of status and pressures.* EEA Report No 8/2012. https://www.eea.europa.eu/publications/european-waters-assessment-2012

Halliday, S. (2001). *The great stink of London.* Dublin: History Press.

Heugens, E. H. W., Hendriks, A. J., Dekker, T., van Straalen, N. M., & Admiraal, W. (2001). A review of the effects of multiple stressors on aquatic organisms and analysis of uncertainty factors for use in risk assessment. *Critical Reviews in Toxicology, 31*(3), 247–284.

Keiser, D. A., & Shapiro, J. S. (2018). *Consequences of the Clean Water Act and the demand for water quality.* No. NBER Working Paper No. 23070. Cambridge, MA. http://www.nber.org/papers/w23070.pdf

Kelly, V., Stets, E. G., & Crawford, C. (2015). Long-term changes in nitrate conditions over the 20th century in two Midwestern Corn Belt streams. *Journal of Hydrology, 525*, 559–571.

Lassaletta, L., Billen, G., Garnier, J., Bouwman, L., Velazquez, E., et al. (2016). Nitrogen use in the global food system: Past trends and future trajectories of agronomic performance, pollution, trade, and dietary demand. *Environmental Research Letters, 11*(9), 095007.

Loew, A., Jaramillo, P., & Zhai, H. (2016). Marginal costs of water savings from cooling system retrofits: A case study for Texas power plants. *Environmental Research Letters, 11*(10), 104004.

Madden, N., Lewis, A., & Davis, M. (2013). Thermal effluent from the power sector: An analysis of once-through cooling system impacts on surface water temperature. *Environmental Research Letters, 8*(3), 35006.

Meybeck, M., Lestel, L., Carré, C., Bouleau, G., Garnier, J., & Mouchel, J. M. (2018). Trajectories of river chemical quality issues over the

Longue Durée: The Seine River (1900S–2010). *Environmental Science and Pollution Research, 25*(24), 23468–23484.

Miara, A., Vörösmarty, C. J., Macknick, J. E., Tidwell, V. C., Fekete, B., et al. (2018). Thermal pollution impacts on rivers and power supply in the Mississippi River watershed. *Environmental Research Letters, 13*(3), 34033.

Mintz, E., Bartram, J., Lochery, P., & Wegelin, M. (2001). Not just a drop in the bucket: Expanding access to point-of-use water treatment systems. *American Journal of Public Health, 91*(10), 1565–1570.

Raptis, C. E., Boucher, J. M., & Pfister, S. (2017). Assessing the environmental impacts of freshwater thermal pollution from global power generation in LCA. *Science of the Total Environment, 580*, 1014–1026.

Raptis, C. E., & Pfister, S. (2016). Global freshwater thermal emissions from steam-electric power plants with once-through cooling systems. *Energy, 97*, 46–57.

Schullehner, J., Hansen, B., Thygesen, M., Pedersen, C. B., & Sigsgaard, T. (2018). Nitrate in drinking water and colorectal cancer risk: A nationwide population-based cohort study. *International Journal of Cancer, 143*(1), 73–79.

Seitzinger, S. P., Mayorga, E., Bouwman, A. F., Kroeze, C., Beusen, A. H. W., et al. (2010). Global river nutrient export: A scenario analysis of past and future trends. *Global Biogeochemical Cycles, 24*(2), GB0A08.

Sinha, E., Michalak, A. M., & Balaji, V. (2017). Eutrophication will increase during the 21st century as a result of precipitation changes. *Science, 357*(6349), 405–408.

Snyder, C. S. (2017). Enhanced nitrogen fertiliser technologies support the "4R" concept to optimise crop production and minimise environmental losses. *Soil Research, 55*(5–6), 463–472.

Stets, E. G., Kelly, V. J., & Crawford, C. G. (2015). Regional and temporal differences in nitrate trends discerned from long-term water quality monitoring data. *Journal of the American Water Resources Association, 51*(5), 1394–1407.

Thornton, P. K. (2010). Livestock production: Recent trends, future prospects. *Philosophical Transactions of the Royal Society B: Biological Sciences, 365*(1554), 2853–2867.

Van Meter, K. J., Van Cappellen, P., & Basu, N. B. (2018). Legacy nitrogen may prevent achievement of water quality goals in the Gulf of Mexico. *Science, 360*(6387), 427–430.

Veolia and IFPRI. (2015). *The murky future of global water quality.* Washington, DC: International Food Policy Research Institute. http://www.veolianorthamerica.com/sites/g/files/dvc596/f/assets/documents/2015/04/IFPRI_Veolia_H2OQual_WP.pdf

Wen, Y., Schoups, G., & van de Giesen, N. (2017). Organic pollution of rivers: Combined threats of urbanization, livestock farming and global climate change. *Scientific Reports, 7*, 43289.

IV

Water Resources Supply and Demand in Context

CHAPTER 13

Opportunities for Water Management

13.1 Review of Water Supply and Demand in a Changing World

People make use of water for many purposes, including water supply for irrigation, for energy production, for drinking, for recreation, and for **ecosystem services**. Although water resources are renewable, they are not inexhaustible. Throughout history, civilizations have flourished with abundant water, accomplished engineering feats to secure its presence, and collapsed due to the lack thereof. Today, human influences on the environment are even greater than they were in the past, which poses major challenges. Planning for water resources must incorporate knowledge of the spatial and temporal availability of water for various uses. The first step in assessing how water challenges can be met is to learn how much fresh water exists on Earth, where it is stored, how it moves, and how it is used.

The general circulation of water through the atmosphere, oceans, rivers, lakes, and beneath the ground determines the availability of fresh water resources at any location on Earth. The motion of water can be described at many different scales. The fundamental concept of hydrology is the **hydrological cycle**—the global-scale, endless recirculatory process linking water in the atmosphere, on the continents, and in the oceans. The atmospheric heat engine drives, and itself is affected by, precipitation and evaporation, leading to broad climate zones. Precipitation and evaporation link the land surface and the atmosphere.

Water availability varies geographically, seasonally, and across longer time periods. Over 80% of water used by humans is from freshwater surface sources. Rivers are the single greatest source of freshwater withdrawals, and the construction of reservoirs on rivers has had a marked impact on river flows and water storage. Flows and storages of continental surface water are determined by topography, geology, and land surface characteristics.

Although surface water is most commonly used and stored by humans, **groundwater** is the largest accessible source of unfrozen, fresh water by volume. The geological properties of rocks control the storage and flow of **groundwater**. Flow is driven by recharge provided by **infiltration** of water at the land surface and by topography. **Groundwater** is available as a water resource either through springs or from wells constructed to access **groundwater**. Overdrafts of **groundwater** lead to depletion and to impacts such as land subsidence. Because **groundwater** can supply discharge to streams, rivers, and springs, **groundwater** depletion can affect surface water resources and ecosystems. **Groundwater** is also tightly linked with soil water; water held in soils is in the unsaturated zone, above the **water table**. Although water held in soils cannot be withdrawn for human uses, soil water is fundamental to plants. Water in the rooting zone is available to plants. Water below the rooting zone is a source of **groundwater** recharge.

Water demand, like water supply (Box 13.1), varies both spatially and temporally, so there is a need to understand the context of water demand both by scale and by user (Box 13.2). Water links together food and energy, and it is central to sustainable economies in cities and reliable resource supplies for domestic uses. Equally important, but often overlooked, is the value of water for environmental uses and **ecosystem services**.

More than 70% of human water use is for agriculture globally, but in the least developed countries of the world, water demand for agriculture can be higher than 90%. Although the predominant agricultural production system in the world is rainfed agriculture, irrigation can double agricultural production and is adopted increasingly. Managing water for agriculture will be vital in supporting rural development and economic growth over the next century, but so will managing water for energy.

Globally, 15% of the world's water withdrawal is used for energy production, but this value varies spatially and temporally and depends highly on the energy technology. Water is withdrawn and consumed to extract, transport, and process primary energy sources like natural gas and coal, but often the largest fraction of water withdrawn and consumed for energy is during electricity production.

Box 13.1 Water Availability in the ACF

The Apalachicola-Chattahoochee-Flint River Basin (ACF) in the southeastern part of the United States is an important municipal, industrial, and agricultural area with diverse ecosystems including a commercial oyster fishery. The necessary water for people, crops, and ecosystems is supplied by the rivers and groundwater aquifers in the basin.

The Flint and Chattahoochee rivers flow from northern Georgia and meet at Lake Seminole to form the Apalachicola River, which flows south into the Apalachicola Bay (Figure B13.1.1). The ACF spans an area of approximately 50,000 square kilometers across eastern Alabama, western Georgia, and the Florida Panhandle.

Figure B13.1.1 Map of the Apalachicola-Chattahoochee-Flint River Basin.

The Chattahoochee River, which is approximately 440 miles long, rises in the foothills of the Blue Ridge Mountains and flows into Lake Sidney Lanier (also called Lake Lanier). The city of Atlanta is located approximately 80 km south of Lake Lanier, which is Atlanta's main water supply. The Chattahoochee River then meanders another approximately 550 km before it reaches Lake Seminole. Between Lake Lanier and Lake Seminole there are other notable impoundments, one of which is West Point Lake.

To the east of the Chattahoochee River Basin lies the Flint River Basin, which begins just south of Hartsfield International Airport serving the metropolitan Atlanta area. The Flint River flows approximately 555 km south, where it reaches Lake Seminole and mixes with the water from the Chattahoochee River.

Flow out of Lake Seminole forms the Apalachicola River, which runs 170 km south before entering Apalachicola Bay. Apalachicola Bay is an estuary—a mixture of fresh and salt water—and is protected from the open Gulf of Mexico by a series of barrier islands.

The Upper Floridan Aquifer is a major aquifer. This aquifer is an important source of water to the Lower Flint and Lower Chattahoochee rivers, as these rivers and their tributaries cut directly into the aquifer in many places. Groundwater in the Upper Floridan Aquifer discharges directly into these rivers. The Upper Floridan Aquifer is a major source of irrigation water in the lower Flint River Basin. The Clayton and Claiborne aquifers are major sources of water in the middle Flint and Chattahoochee basins. Other aquifers in the region supply smaller amounts of water for individual domestic use and for irrigation in other parts of the ACF.

Box 13.2 Water Use in the ACF

The major anthropogenic demands for water in the Apalachicola-Chattahoochee-Flint River Basin are for cities and industries, for electricity production (thermoelectric power plants), and for agriculture. There also are needs for environmental flows, particularly downstream in the Apalachicola River where the river flows from Georgia into Florida. Although the region is, in general, well-endowed in terms of water resources, population growth and the expansion of irrigated agriculture have resulted in considerable stress, leading to the need manage water resources to balance competing demands.

The Atlanta metropolitan area is at the northern end of the ACF (see Figure B13.1.1). This area had a population of about 1.7 million in 1970 and grew to a population of about 5.3 million by 2010. Future projections place the expected population by 2050 at about 8 million (Black & Veatch, 2017). The main water supply for the area is Lake Sidney Lanier, which was impounded in 1950. Total water withdrawals for the metropolitan area in 2014 amounted to about 20,000 cubic meters (m^3)/day; by 2050, the demand is projected to grow to perhaps 28,000 m^3/day (Black & Veatch, 2017). Much of this water is not used consumptively; that is, much of the water is returned to downstream flows as wastewater. If 25% of the water withdrawn is consumed, the consumptive use by the Atlanta metropolitan area in 2014 would have been 5,000 m^3/day, and the projected consumptive use in 2050 would be 7,000 m^3/day.

There are a number of thermoelectric power plants that withdraw water from ACF rivers. For once-through cooling plants, much of the water is returned to the river, albeit at a higher temperature than when it was withdrawn. There is some consumptive use with once-through plants; plants that use cooling towers have even higher consumptive uses. The estimate for consumptive use by ACF plants in 2015 was about 174,000 m^3/day (Faeth et al., 2018). Electricity demand in Georgia may increase by about 30% by 2050 (Faeth et al., 2018), so consumptive demands for the ACF could increase proportionally.

As with almost all regions of the world, consumptive water use in the ACF is dominated by irrigated agriculture. In the ACF, a large fraction of irrigation withdrawals is in the Flint River Basin, with the bulk being pumped from the Upper Floridan Aquifer. According to estimates, the irrigated area of cropland in Georgia increased tenfold from about 58,000 hectares in 1970 to about 580,000 hectares in 2008 (University of Georgia, 2009) and remained relatively steady at that level for the previous decade. About half of the irrigated area is in the Flint River Basin (Banerjee et al., 2007). Assuming a total irrigation depth of 0.38 m (e.g., Torak & Painter, 2011) for an area of 290,000 hectares, the water withdrawn for irrigation would be about 3 million m^3/day. If the irrigation efficiency is 0.75, the consumptive use would be about 2.2 million m^3/day.

These water demands are based on annual averages. The actual demand is much greater in summer months when streamflow is naturally low (e.g., see Figure 2.6). The impact of large water withdrawals and consumptive use is most pronounced at times of low flow or droughts. There have been notable declines in low flows in the Flint River as a result of groundwater withdrawals (see Box 3.2) for irrigation (Rugel et al., 2012).

Riverine and floodplain ecosystems are sensitive to low flows, so upstream consumptive uses in the ACF can impact ecosystems in the Apalachicola River Basin in Florida. The Apalachicola has extensive Tupelo cypress swamps and a large number of fish and mussel species. The Apalachicola River discharges into the Gulf of Mexico at the Apalachicola estuary, which is an important oyster fishery in addition to hosting numerous fish species that depend on a mix of fresh water and salt water and are thus sensitive to declining low flows. This points out the need to manage water resources to balance the trade-offs among competing uses of water in a way that meets ecological low-flow requirements, as discussed in Chapter 8.

In developed countries, water withdrawals for **thermoelectric** power can be greater than water withdrawals for irrigation. In the coming century, **anthropogenic** climate change will increase the pressure on society to develop low-carbon energy technologies, some of which may have high water-use intensities.

Domestic water use is the most obvious source of water demand to individuals—water is required for drinking, cooking, cleaning, and bathing. Domestic water use in cities varies tremendously across the world, driven by a host of climate and socioeconomic factors. As urbanization increases, stresses are placed on supply systems that reflect population growth and economic development.

Whereas domestic water use is the most obvious source of water demand to individuals, the most obscure demand is often water for environmental purposes. Environmental water uses, such as water to support wetlands, are difficult to calculate, controversial to monetize, and not understood well by the public. Although environmental water use is not often in direct competition with other demand-side sectors, environmental water allocations can create controversy when water is scarce.

Understanding how water is currently used provides a basis for understanding future changes. We are in an era of change that has never been seen before on Earth. Human population is growing rapidly and is more mobile, consumption of our natural resources is increasing and often outpaces the rate of renewal or production, and climate change is transforming our planet. The effect of increases of greenhouse gases in the atmosphere is to warm the lower atmosphere, leading to increased intensity of the **hydrological cycle**. Exactly what the impacts will be at any location are uncertain, but broad impacts will greatly affect water resources and our environment as well as food and energy production.

Water resource management and the formulation of sound policy are important as we move into a generation with significant growth in population and development. Management decisions often involve balancing conflicting demands for water (Box 13.3). There are many means of formulating policy and effectuating management of water resources, but water law often sticks out, especially in the United States. The goal of law is to avert disputes or resolve conflicts by formally identifying standards and expectations that

Box 13.3 Water Management in the ACF

The complex and difficult decisions about water management in the Apalachicola-Chattahoochee-Flint River Basin (ACF) involve the potential conflicting needs of supplying municipal water (principally to the Atlanta metropolitan region), industrial water (especially to thermoelectric power plants), and irrigation water in Georgia versus maintaining flows to sustain ecosystems during summer in the Apalachicola in Florida. To gain perspective on comparing the water demands in Georgia with the ecological low-flow requirements (Chapter 8) in Florida, let us first convert the average consumptive use in cubic meters per day to the average flows in cubic meters per second. The estimated consumptive demand for the Atlanta metropolitan region is equivalent to about 0.1 m^3s^{-1}, and the thermoelectric consumptive demand is about 2 m^3s^{-1}. The consumptive use by irrigation averages about 25 m^3s^{-1}. This is an average over a year and does not take into account that irrigation occurs primarily in a few months in the summer. We might expect that in critical months the consumptive irrigation demand would exceed 100 m^3s^{-1}.

The key management trade-off for the ACF is the irrigation water withdrawals in Georgia versus ecological low-flow requirements in Florida. There are a number of flow requirements that have been proposed, among them that daily low flows should not be below a certain number of days per year (Richter et al., 2003). For the Apalachicola, one such criterion is limiting the number of days when flow falls below 155 m^3s^{-1} to a maximum of 24. This threshold criterion has been violated much more frequently since 1970 when irrigation in the Flint River Basin started to increase (Figure B13.3.1). The evidence is rather clear that consumptive water use in Georgia has

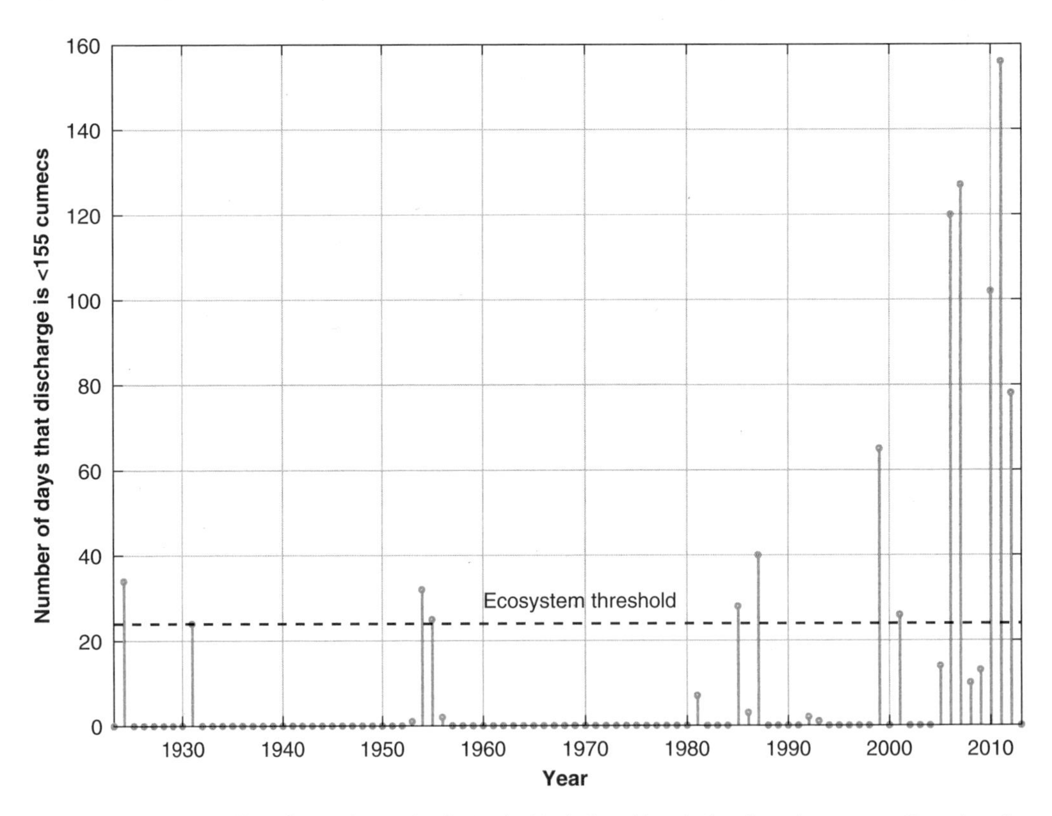

Figure B13.3.1 Ecological low-flow violations for the Apalachicola from historic data from the gauge at Chattahoochee, Florida, where the discharge from Georgia into the Apalachicola in Florida occurs. Data from the US Geological Survey.

negatively impacted low flows that affect ecosystems in the Apalachicola and its floodplains.

Management of scarce resources involves trade-offs, which is the case when multiple political jurisdictions share a resource. A mutual agreement must be reached among Georgia, Alabama, and Florida in this case on how to manage water resources in the ACF. The ACF Stakeholders group, an independent organization formed to advise about planning, recommended a set of actions to conserve water (ACF Stakeholders, 2015). Many steps already have been taken to conserve water in the basin. For example, water withdrawals for the Atlanta metropolitan region have declined in the twenty-first century even while the population has been increasing. Other water use efficiency measures have been taken by industry. The ACF Stakeholders group recommended a reduction of consumptive use of water by agriculture; although some efficiencies have been gained through irrigation technologies, the consumption due to irrigated agriculture has not diminished.

Water management during even moderate drought conditions is especially important, as it is under these conditions that ecological low-flow requirements are not met. ACF Stakeholders (2015) recommended that "the states of Alabama, Florida, and Georgia should collaborate in the development plan." The stakeholder group noted that "drought management involves temporary equitable reductions in water uses during droughts to conserve water so that deeper reductions or even catastrophic shortages can be avoided." Unfortunately, the history of interstate water agreements in the ACF is not rife with examples of serious collaboration (Ruhl, 2005). The latest suit brought by Florida seeking to obtain more water to maintain ecological flow requirements was sent back to a Special Master by the US Supreme Court in late June 2018 to determine a reasonable plan for equitable apportionment of water in the ACF among the states.

are enforceable by penalties against violators. Nevertheless, the relationship, or lack thereof, between legal systems and water resources can complicate **water scarcity** and make it difficult to manage water resources sustainably and equitably.

In countries where water resources are governed by customary laws and economies are transitioning, there may be limited resources or limited political capital to devote to developing and enforcing customary laws. Realistically, a combination of institutional and behavioral adaptations, as well as changes in physical **infrastructure**, will be required to achieve sustainability. Although there are no panaceas in **water management**, there are many opportunities to manage water sustainably in the future.

13.2 Water Management in Context

The management of water resources is aimed at reliably supplying water of good quality to satisfy the demands of different uses to as great an extent as possible and in as fair and equitable a manner as possible. This includes consideration of environmental flows and **ecosystem services**.

Water management is distinct from **water governance**: "management refers to activities of analyzing and monitoring, developing and implementing measures to keep the state of a water resource within desirable boundaries . . . governance sets the rules under which management operates" (Pahl-Wostl et al., 2012). These rules can be set formally and informally. To be most successful, the rules should account for multiple stakeholder interests. Some examples of governance systems include government and nongovernment organizations, property rights to water, operational rules, collective-choice rules, monitoring processes, and sanctioning processes (Ostrom, 2009). Although the definition of water resource governance varies, three key concepts tend to be included in the definition: (1) decision-making processes (2) through institutions (3) involving multiple actors. **Water governance** defines the water resource goals and management practices, while **water management** is the action of meeting the defined goals through the identified management practices and the associated outcomes (Lautze et al., 2011).

With regard to **water management**, there are two broad approaches to match supply and demand. One is to avoid shortages in time by storing water during times of plenty to be used during dry periods or to avoid shortages at one location by importing water from elsewhere. The other is to take steps to lessen the amount of water used for various purposes. Broadly speaking, the two management approaches are to (1) increase supplies and (2) decrease demands.

Augmenting supplies for a location involves the development of **infrastructure**, often requiring large projects and associated investments. Surface water supplies are developed by constructing dams, canals, pipelines, pumping stations, and other associated **infrastructure**. These developments can be to supply water for a variety of purposes including water for municipalities, for irrigation, for hydropower, and for recreation. As population of a region expands, increasingly more ambitious projects are taken on. One case that exemplifies this expansion of **infrastructure** projects through time comes from Los Angeles, California (Box 13.4).

Groundwater supplies are expanded by constructing additional or deeper wells. **Groundwater** storage also can be enhanced by constructing **infiltration** ponds to allow storm water or recycled

Box 13.4 Water Supply Expansion in Los Angeles, California

In the second half of the nineteenth century, Los Angeles began a period of rapid growth. Initially, the city used the local surface water (e.g., the Los Angeles River) and diverted these waters for municipal use and for irrigation. The groundwater in the Los Angeles Basin also was tapped to supply the growing demands. It soon became clear to planners who wanted to spur urban and agricultural growth, however, that water from outside the region would have to be imported.

Surreptitiously water rights in the Owens Valley on the eastern side of the Sierra Nevada Mountains were purchased. An aqueduct several hundred kilometers in length was constructed with the necessary siphons to allow water to flow by gravity to Los Angeles. At the time the aqueduct was constructed, it was the largest water project ever undertaken. The aqueduct opened in 1913 and led to the infamous "Owen Valley Water Wars." Nevertheless, even this herculean effort was not enough.

In 1922, California gained water appropriations from the Colorado River, and in 1928 the construction of Hoover Dam was authorized. The aqueduct from the Colorado River to Los Angeles stretches almost 400 km across the Mojave Desert and includes canals, tunnels, siphons, and pumping stations to get the water over mountain ranges. In 1960, the State Water Project was started to transport water from the relatively wet northern part of California to the drier southwestern part of California. The project included the construction of 21 dams and over 1,000 km of pipelines, canals, and tunnels. The construction of all these massive engineering projects to increase the supply of water allowed the explosive growth of the Los Angeles area in the twentieth century.

wastewater to recharge aquifers. For example, Los Angeles has used **infiltration** ponds to recharge the **aquifer** with a mixture of excess river water and wastewater treated to drinking water standards. In recent years, managed **aquifer** recharge has expanded throughout the basin to make more beneficial use of storm runoff waters from the area rather than allow the water to discharge to the ocean.

The other major category to manage water resources, limiting the demand for water, can be done through a variety of measures. Cities have numerous options to stimulate **water conservation** measures. In the United States, for example, almost 80 different **water conservation** policies have been identified (Hess, Wold, Worland, et al., 2016). These include a number of metrics on requirements to reduce water use under prescribed circumstances, on the award of rebates to customers who install efficient appliances, and on pricing of water to encourage conservation (Table 13.1).

As cities progress and the options for increasing water supplies diminish, ideally a transition to a **water conservation** approach occurs. Sociotechnological systems often undergo large shifts in operation as the system matures. Such shifts have been called *regime transitions*. An ideal typical trajectory of three phases of water-supply regime transition can be identified (Hess, Wold,

Table 13.1 Selected measures to encourage water conservation.

Limits	Watering lawns and gardens permitted only during certain times and certain days of the week Washing driveways or sidewalks limited to certain times, certain days, or for sanitation purposes
Requirements	Leaks, breaks, or malfunctions in the home must be fixed New or retrofitted construction must • have water-efficient toilets, faucets, showerheads, etc. • have water-efficient dishwashers, washing machines, etc. • have water-efficient irrigation, such as a rain-sensing sprinkler system • be landscaped with plants and trees that reduce or eliminate the need for supplemental water from irrigation
Rebates	Rebates are provided for • purchasing a water-efficient toilet, faucet, showerhead, etc. • purchasing a water-efficient dishwasher, washing machine, etc. • purchasing products that improve the water efficiency of pools, such as a pool cover • purchasing and installing water-efficient irrigation, such as a rain-sensing sprinkler system • landscaping with plants and trees that reduce or eliminate the need for supplemental water from irrigation • setting up a rainwater-harvesting system • setting up a graywater-recycling system

Hunter, et al., 2016). The first phase is to use local **groundwater** and surface water sources while the population and demand are low. As an urban center grows and imposes ever greater demands for water, a second phase is the development of long-distance pipelines and canals from distant reservoirs that import water from elsewhere (Perrone et al., 2011). After urban centers exhaust local and long-distance sources of water, they enter a third phase—(1) the development of alternative technological methods such as managed **aquifer** recharge and desalination and (2) increased emphasis on nontechnological methods to reduce demand for water. For example, Los Angeles has adopted more and more conservation measures and expanded its managed **aquifer** recharge program to limit its present and future requirements to use water from beyond its borders (Box 13.5).

The two largest water-use sectors are agriculture and energy, so large gains in efficiencies can make a significant impact on reducing demands to meet supplies during times when water is scarce. In agriculture, choosing crops according to water needs and installing the most technically efficient irrigation systems can reduce the demand for water. As consumers, we can purchase food that is more water efficient or purchase food from local farmers

Box 13.5 Water Conservation Measures in Los Angeles, California

Los Angeles experienced the first two phases of the ideal water supply regime transitions—importing water from afar and improving efficiency of water use through technology—from the latter half of the nineteenth century through much of the twentieth century. In recent decades, the move to the third phase of regime transition—nontechnological approaches to reducing demand—has taken place. Los Angeles has many policies and regulations in place to encourage water conservation. Even in comparison with other cities in the arid and semiarid southwestern region of the United States, Los Angeles is a leader in the number of measures that they have taken (Hess et al., 2016).

The Los Angeles Department of Water and Power (LADWP) has an ongoing program to manage water demand, a program that includes public education and outreach, conservation water pricing, and an ordinance that prohibits waste of water (Los Angeles County Waterworks District 29, 2017). The program also offers a set of rebate options for replacement of toilets, recirculating hot water pumps, clothes washers, turf grass, irrigation controllers, pool covers, weather-based irrigation controllers, and rain sensors. In addition to the program elements in effect for all years, the program also includes a phased approach for instituting mandatory restrictions during drought years. The need for and the effectiveness of such mandatory restrictions is evident from the historical information.

Los Angeles suffered three serious consecutive drought years from 2008 to 2010. In 2008, voluntary water restrictions were in effect. In 2009, mandatory restrictions such as limitations on the frequency and duration of lawn watering were put in place. In 2010, the mandatory restrictions were made more stringent: the allocation of water to each household was reduced by 15%. The price of water beyond the reduced allotment was increased by 44%. The voluntary restrictions did not have an apparent effect on water use. The mandatory restrictions in 2009 achieved some savings, and the extended mandatory restrictions and price structure change in 2010 led to a 23% decrease in water use (Mini et al., 2015).

To reduce the dependence on water imported from long distances, the reliability of which is expected to decline in the face of changes in climate, Los Angeles plans to extend the use of managed aquifer recharge of captured stormwater runoff. In 2015, LADWP induced about 100 million liters per day of collected storm water to recharge groundwater, and this could be extended over the period to 2035 to 230 million to 390 million liters per day (Geosyntec Consultants, 2015). The current recharge represents about 6% of total current demand, and the projected future amounts represent 14% to 24% of demand.

who have adopted water-efficient technologies. In energy, adopting dry cooling technologies for thermoelectric power generation can reduce water withdrawals and consumption. Moving away completely from thermoelectric power toward the use of technologies that use renewable resources such as solar photovoltaic (PV) or wind can drastically cut water demands. As consumers, we can reduce our energy demands, which in turn, reduce our water demands.

The use of water resources is linked with the use of plentiful food and energy, both fundamental to our modern-day lifestyles. In addition to the technological and nontechnological measures mentioned above that apply directly to water resources alone, there are management options for other resources that can impact water (Perrone & Hornberger, 2014). In effect, reductions in energy use, food waste, and consumption of foods with high water requirements would reduce demands for water. In the future, it is very likely that success in managing water resources will be linked to the integration of land and energy resources management and governance systems.

13.3 Water Security

Water security has quantity and quality components. In this context, the terms *availability* and *reliability* relate to the physical security of the resources and are an indication of a sufficient quality (e.g., water without bacteria or viruses if for human uses) of a minimal quantity. Recent works have expanded the definition of water security to acknowledge risks to the economy or environment from water or the lack thereof (Grey & Sadoff, 2007). *Water insecurity* can be transitory or chronic. *Transitory* insecurity is short term, which can be a function of political instability or extreme weather events. *Chronic* insecurity is long term. Chronic insecurity often stems from overallocation of resources, lack of investments in **infrastructure**, or the disregard for environmental externalities, which can affect **ecosystem services**.

Water security can be impacted by pollution, which can also lead to transitory or chronic insecurity of fresh water suitable for human consumption. As we know from Chapter 1, residence times for **groundwater** tend to be orders of magnitude greater than for surface water. Thus, contamination of **groundwater** and surface water do not necessarily have the same impact on water security. While surface water is generally more vulnerable to contamination than **groundwater**, flushing pollution from **groundwater** reservoirs tends to take much longer than from surface water reservoirs. One opportunity to manage water sustainably is to protect the freshwater resources that are readily available to us (Box 13.6). Water quality and water quantity are interrelated. Even places with abundant water resources can feel the impacts of limited water supplies if the supplies are not of a suitable water quality for the uses in question.

Box 13.6 Management to Control Nutrient Pollution

Conventional agriculture, widely practiced throughout the world, relies on heavy inputs of nutrients and other agricultural chemicals (Chapter 5). These chemicals can be transported to surface water and groundwater and lead to significant contamination issues. For example, in the United States, some 11 billion kilograms of nitrogen fertilizer are applied on croplands each year with up to half of it discharged to streams by either surface runoff or groundwater flow (Capel et al., 2018). It has proven to be quite difficult to manage the runoff of nitrogen from agricultural fields, but it is essential that these fluxes be controlled (Chapter 12).

In the 1990s, a study by the US Geological Survey in the Apalachicola-Chattahoochee-Flint River Basin indicated that the presence of extensive wetlands and forests in the floodplains of streams and rivers acted as protection from nitrogen pollution. Nevertheless, the subsurface drain waters in tile-drained fields had very high nitrate concentrations and represented a pathway for contaminant movement to streams (Frick et al., 1998).

Maintaining or reestablishing riparian buffer zones is one management option for controlling nutrient runoff. Others are to use conservation tillage (Chapter 5), to minimize surface runoff from irrigation, to decrease the use of agricultural chemicals, and to use technologies such as constructed wetlands to treat tile drain waters (Chapter 8). All these practices are generally useful, but it should be noted that if the agricultural chemical of concern is moving primarily through groundwater, conservation tillage and forest buffer zones can be somewhat counterproductive in that they increase infiltration rates (Capel et al., 2018).

13.4 Water Scarcity

Although scarcity can be a simple concept—supply does not meet demand—it is complex in its application to water resources. Traditionally, the term **water scarcity** is used mostly from a water quantity perspective. Although **water scarcity** is mostly applied to water quantity, contamination of water resources can impact or limit its use; the nexus between water quantity and quality becomes especially visible as competition for water increases. Scarcity can be defined by the absolute measure of the amount of water (i.e., *physical* **water scarcity**), but also it can be defined by a lack of investment in water **infrastructure** or political, cultural, and social constructs (i.e., *institutional* **water scarcity**) (International Water Management Institute, 2007).

Physical **water scarcity** is often associated with arid regions, but this may not be the case always. Arid regions with few water demands may not be water scarce whereas humid regions that have overallocated their resources may be water scarce. Physical **water scarcity** can impact the environment, especially when water governance prioritizes human uses only. Physical **water scarcity** can also occur on transitory timescales. Drought can suddenly and drastically reduce available water supplies, increasing the competition for water resources and challenging status quo governance and management.

Institutional **water scarcity** is associated with areas that may have sufficient water resources under effective governance and management but are limited by institutional resources, such as current legal frameworks, markets, or **infrastructure** for the conveyance and delivery of potable water (National Research Council, 2001). Managing **water scarcity** in a sustainable manner will require a diverse portfolio of tools customized to the needs of each place, yet adaptable to changes over time as regions develop and change.

13.5 Competition for Water and Trade-offs

Resolving the conflicts surrounding the increasing competition for fresh water can be viewed as an optimization problem of allocating scarce resources to maximize utility. More and more, resource managers are recognizing that solving water problems successfully requires the addition of social, institutional, cultural, and legal contexts. As the competition for fresh water increases, so does the focus on preserving water quality so that it is suitable for human consumption and the environment. It is easy to be pessimistic about the path forward, given all of the constraints placed on our water resources. Nevertheless, technological advancements are being made at a pace never seen before, and the opportunities for increasing efficiency in our water systems and water treatment facilities also are greater than humans have ever seen before. An optimistic view of the path forward is that we are transitioning to an era where technical developments and forward-looking development of governance—the social, economic, and administrative systems that influence management—will jointly lead to success in managing water resources sustainably.

Much of the **infrastructure**, institutions, and laws and **regulations** that constitute the framework for determining how water is used were developed decades ago in many developed countries like the United States. In some instances, more water has been allocated for different uses than is actually available, especially dur-

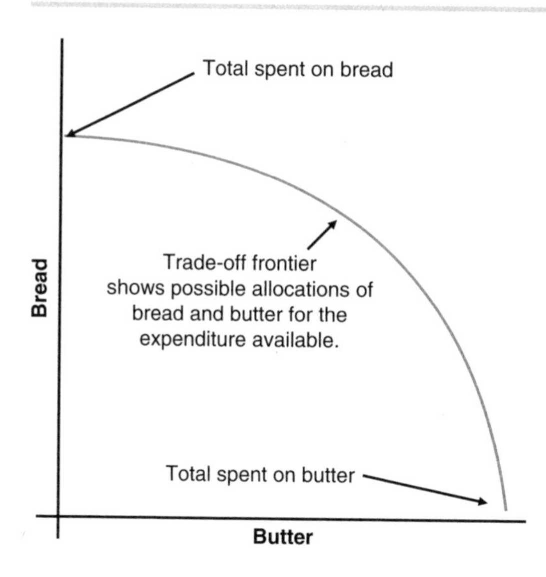

Figure 13.1 A simple representation of a trade-off frontier. Given a fixed sum of money to spend, the total amount could be spent to buy butter, the total amount could be spent to buy bread, or the amount could be split between butter and bread with a range of possible allocations between the two commodities.

ing dry periods. This has led to crises of overdrafts of both surface and **groundwater** with detrimental impacts on the environment. As demands for water continue to increase, we will see instances of **water scarcity** arise more frequently. Without proper safeguards for water quality, the amount of water for some uses may be further limited. This means that choices, often very difficult choices, must be made about what uses receive what fractional part of their demand. Some uses will be met more fully than others while others will be met only partially or even with minimal attention. That is, the choices involve trade-offs between and among the various uses of water.

A convenient way to visualize management options for difficult choices is a **trade-off frontier**, a curve that depicts the benefits associated with different mixes of allocations. Economists often consider trade-offs in terms of expenditures. For example, suppose you have a fixed amount of money to spend and the choice is to purchase bread or butter or some combination of both. The **trade-off frontier** would be drawn on a graph with "bread" as one axis and "butter" as the other. One point on the frontier could be where all money is spent on bread and none on butter. Another point could be all butter and no bread. The **trade-off frontier** is the curve that shows all of the possible combinations (Figure 13.1). Note that a decision about what allocation set to choose is not determined by the calculation of the frontier—in the simplistic example in the figure, the choice is up to the consumer about what mix of bread and butter is desirable, given the fixed amount of money available to spend. The frontier itself only informs the decision maker about all the possible combinations that use fully the available resources; the combinations that do not lie on the frontier are either not feasible (i.e., points above the curve) or inefficient (i.e., points below the curve).

A **trade-off frontier** for water allocation follows a similar logic except the "expenditure" is in terms of water and not money. Under scarcity, a fixed amount of water must be allocated among competing uses. One can get so much irrigation or recreation or municipal supply or hydropower or other items with the amount of water available. Similar to the simple example of bread and butter, the calculation of how much benefit can be derived from various allocations of water does not result in the identification of the "best" management option, only the possible combinations that use fully the available resources. One example is the trade-off faced in Sri Lanka of how to allocate water between irrigation and hydropower (Figure 13.2). Ultimately the choice of how to allocate water under scarcity must be made by managers who take into account how various stakeholders will value different aspects of water use, including the provision of water for ecosystems or sociocultural aspects of water.

Although a **trade-off frontier** is a conceptually simple way to visualize the implications of different management choices, there are many calculations and evaluations that must be accomplished to construct the curves. Water availability varies through time and at different locations. Knowledge of the water balance for the region being considered is a key component of the analysis. Because water balances are impacted by development, changes in land use and population will require budgets to be updated and reconsidered. Water quality also plays a key role; if the quality is too poor for use, then the quantity available may shrink if proper treatment facilities are not available.

In addition to estimating the total amount of useable water available, the benefits associated with various options must be developed. These may be expressed in terms of economic variables, but most often the consideration is in terms of the product of a given

Box 13.7 Conflicting Water Demands in the Klamath River Basin

The Klamath River Basin spans northern California and southern Oregon. The basin provides significant water to ranchers and is home to the Klamath Tribes, which have a strong cultural and spiritual connection to the river system. Fundamental to the Klamath Tribes' culture are the native fish within the Klamath River and its tributaries. The Klamath Tribes have long relied on salmon as a staple food for their traditional diet, and salmon have been fundamental to their religion and economy. Many traditional stories surround salmon and fishing, more broadly (US Bureau of Reclamation, 2012).

Between 1864 and 1986, the Klamath Tribes endured a long, unsettling history regarding their land and river resources, representing one of the sorrier chapters of US history (Cohen, 1945). Examples abound, of which the construction of the Copco 1 dam in 1910 provides but one. The Copco 1 dam impeded the migration of salmon in the Upper Klamath Basin. By the 1950s, development within the basin, including increased irrigation and reservoir storage, had led to significant impacts on key salmon and trout species and suckers (*c'waam* and *koptu*) due to low river flows and limited spawning habitat (US Bureau of Reclamation, 2012).

By the 2010s, competition for surface water in the Klamath River Basin was high, even in normal years, due to overallocation of water rights. In 2013, a particularly dry year, water flows in the upper basin were 40% that of normal years. Water was scarce, and to maintain environmental flows, little water could be withdrawn for other purposes. Earlier in the year, the state of Oregon maintained that surface water rights for the Klamath Tribes were dated to "time immemorial," ultimately granting the Klamath Tribes the most senior water rights. Although water rights had not been used historically to maintain environmental flows during dry periods, the Klamath Tribes called their water rights for environmental flows in June 2013. This call curtailed more than 100,000 acres of irrigated nontribal land, fueling the ongoing battle between demand-side sectors within the Klamath River Basin (Learn, 2013).

In May 2018, drought conditions escalated the debate among irrigators, fishers, conservationists, and the tribes regarding water allocation decisions. The Klamath Tribes now rely heavily on suckers as a replacement to their traditional salmon. Drought conditions have threatened suckers because low water levels have led to toxic algae blooms and reduced access to spawning habitat. In response to the low water levels, the Klamath Tribes filed a lawsuit under the federal Endangered Species Act to protect the suckers (Klamath Tribes of Oregon, 2018).

allocation—the fraction of residential demand met, the amount of a crop produced by irrigation, the amount of electricity generated by hydropower, the fraction of time an environmental flow requirement is met, and so forth. Because water may be valued differently by different water users, it is also fundamental to consider cultural or social values (Box 13.7). Albeit an often-difficult task, the construction of a **trade-off frontier** can lead to insights about water allocations.

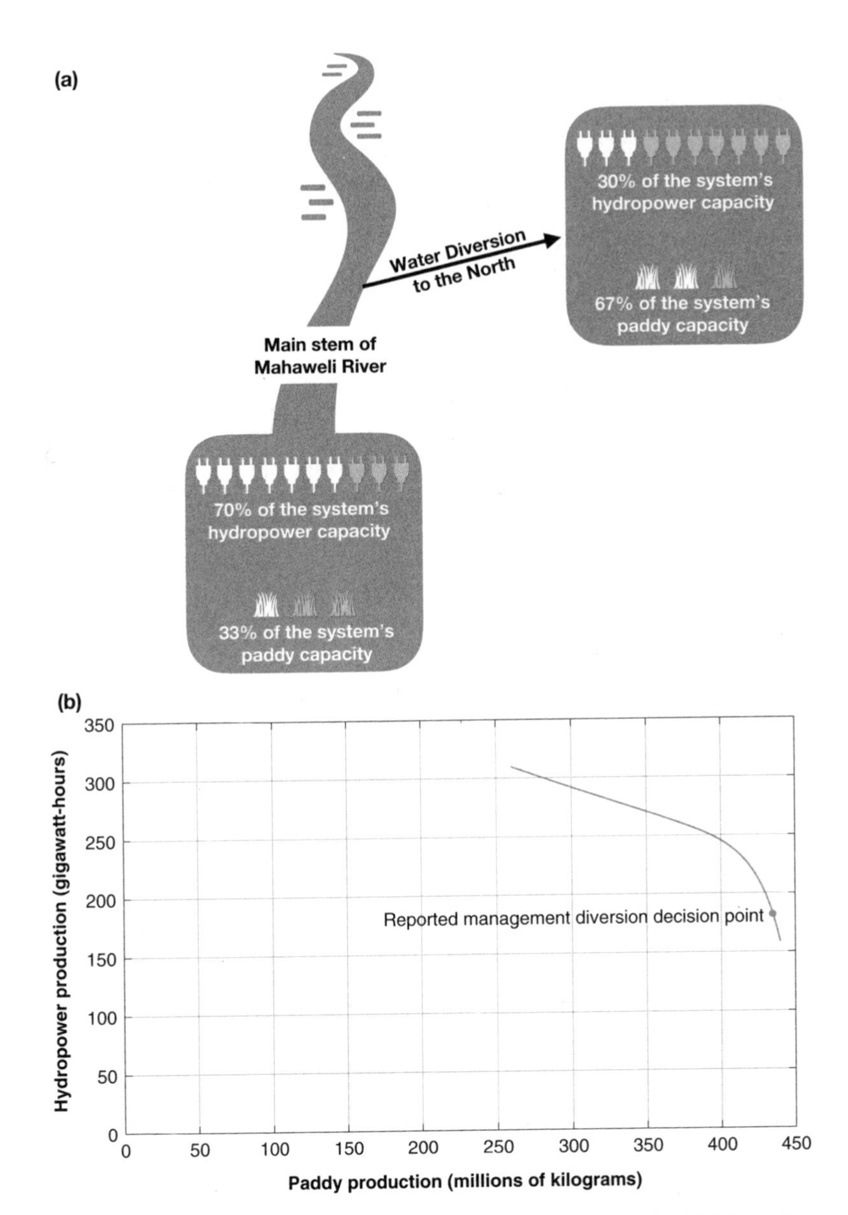

Figure 13.2 Water management in the Mahaweli River Basin in Sri Lanka. (a) Schematic of the decision for allocating water between the main stem of the Mahaweli River for hydropower and the diversion canals to supply paddy fields in the north-central region of Sri Lanka. (b) The trade-off frontier for paddy production and hydropower for a drought year in Si Lanka. The frontier is based on data that relate the amount of water available to paddy and hydropower production. The reported diversion for this drought year shows the strong preference for paddy production even though in monetary terms power is much more valuable than rice. Data from Perrone and Hornberger 2014.

13.6 Room for Optimism

Water resources are highly interconnected with agriculture, energy, cities, and the environment, and are influenced by natural, social, and human-built systems. When water is insufficient to maximize all the demand objectives—either because of low water availability or low water quality—trade-offs are made to allocate water. These trade-offs emphasize the transdisciplinarity of sustainable **water management** and suggest that we need to consider a broad range of systems and valuations when identifying **water management** pathways. The solutions moving us toward sustainable **water management** may indeed be complex, but there is room for optimism.

13.7 Key Points

- Two management approaches are to increase supplies and to decrease demand. (Section 13.2)
- Water security can be transitory or chronic. (Section 13.3)
- Scarcity can be defined by the absolute measure of the amount of water or by a lack of investment in water institutions. (Section 13.4)
- When water is scarce and competition is high, trade-offs between and among various uses of water must be made. (Section 13.5)
- A **trade-off frontier** can be a useful tool to inform decision makers about all the possible combinations that use fully the available resource combinations of paths forward. (Section 13.5)

13.8 Example Problems

Problem 13.1. Choose a country to explore its water resources. Compose a concise paragraph between 50 and 150 words on each of the following:

- An introduction to the selected country, providing information about the country's geography.
- The significance of various water supplies in the selected country.
- The significance of water demands (e.g., domestic, agricultural, energy, environmental) in the selected country.
- The **anthropogenic** drivers of change (e.g., climate change impacts, population stresses, economic growth) in the selected country.
- The opportunities available to address challenges facing the country.
- The trade-offs associated with each of the opportunities outlined.

Problem 13.2. Identify which region your country lies in. Identify two to five classmates with a country in the same region as your country. Provide a synthesis of the state of water and management opportunities in the mutual region. Specifically, address the following:

- Commonalities among the countries' water supplies, demands, and management.
- Differences among the countries' water supplies, demands, and management.
- Current or projected challenges facing the region.
- Opportunities to address the challenges facing the region.
- Trade-offs associated with each of the opportunities outlined.

13.9 References

ACF Stakeholders. (2015). *Sustainable water management plan.* http://acfstakeholders.org/wp-content/uploads/2015/05/ACFS-Sustainable-Water-Management-Plan-For-Release.pdf

Banerjee, S. B., Tareen, I. Y., Gunter, L. F., Bramblett, J., & Wetzstein, M. E. (2007). Forecasting irrigation water demand: A case study on the Flint River Basin in Georgia. *Journal of Agricultural and Applied Economics, 39*(3), 641–655.

Black and Veatch. (2017). *Water resource management plan, Metropolitan North Georgia Water Planning District.* http://northgeorgiawater.org/wp-content/uploads/2018/03/Water-Resource-Management-Plan_REVISED.pdf

Capel, P. D., McCarthy, K. A., Coupe, R. H., Grey, K. M., Amenumey, S. E., et al. (2018). *Agriculture—a river runs through it—the connections between agriculture and water quality.* Circular 1433. Reston, VA: US Geological Survey. https://doi.org/10.3133/cir1433

Cohen, F. S. (1945). *Handbook of federal Indian law.* Washington, DC: US Government Printing Office.

Faeth, P., Hanson, L., Kelly, K., & Rosner, A. (2018). *The water-energy nexus in Georgia: A detailed examination of consumptive water use in the power sector.* Atlanta, GA: Southface Energy Institute/Southern Environmental Law Center. http://www.southface.org/wp-content/uploads/2018/04/Full-Study-The-Water-Energy-Nexus-in-Georgia_April-2018.pdf

Frick, E. A., Hilpe, G. R. B., Couch, C. A., Hopkins, E. H., Wangsness, D. J., & Garrett, J. W. (1998). *Water quality in the Apalachicola–Chattahoochee–Flint River Basin, Georgia, Alabama, and Florida, 1992–95.* Circular 1164. Reston, VA: US Geological Survey. https://doi.org/10.3133/cir1164

Geosyntec Consultants. (2015). *Stormwater capture master plan.* Los Angeles. https://www.ladwp.com/cs/idcplg?IdcService=GET

_FILE&dDocName=OPLADWPCCB421767&RevisionSelectionMethod=LatestReleased
Grey, D., & Sadoff, C. W. (2007). Sink or swim? Water security for growth and development. *Water Policy*, *9*(6), 545–571.
Hess, D. J., Wold, C. A., Hunter, E., Nay, J., Worland, S., Gilligan, J., & Hornberger, G. M. (2016). Drought, risk, and institutional politics in the American Southwest. *Sociological Forum*, *31*, 807–827.
Hess, D. J., Wold, C., Worland, S. C., & Hornberger, G. M. (2016). Measuring urban water conservation policies: Toward a comprehensive index. *Journal of the American Water Resources Association*, *53*(2), 442–455.
Klamath Tribes of Oregon. (2018). Restoring fish and a dying lake . . . The Klamath Tribes: Klamath–Modoc–Yahooskin. http://klamathtribes.org/restoring-fish-and-a-dying-lake/
Lautze, J., da Silva, S., Giordano, M. & Sanford, L. (2011). Putting the cart before the horse: Water governance and IWRM. *Natural Resources Forum*, *35*(1), 1–8.
Learn, S. (2013). Klamath Tribes and federal government put out historic call for water rights in drought-stricken Klamath Basin. *The Oregonian*, June 10. https://www.oregonlive.com/environment/2013/06/klamath_tribes_and_federal_gov.html
Los Angeles County Waterworks District 29. (2017). *2015 Urban water management plan for Los Angeles County Waterworks District 29, Malibu, and the Marina del Rey Water System*. Los Angeles. https://dpw.lacounty.gov/wwd/web/Documents/2015 Urban Water Management Plan for District No. 29 and the Marina del Rey Water System.pdf
Mini, C., Hogue, T. S., & Pincetl, S. (2015). The effectiveness of water conservation measures on summer residential water use in Los Angeles, California. *Resources, Conservation and Recycling*, *94*, 136–145.
Molden, D. (2007). *Water for food, water for life—a comprehensive assessment of water management in agriculture. Comprehensive assessment of water management in agriculture*. London and Sterling, VA: Earthscan with the International Water Management Institute.
National Research Council. (2001). *Envisioning the agenda for water resources research in the twenty-first century*. Washington, DC: National Academies Press.
Ostrom, E. (2009). A general framework for analyzing sustainability of social-ecological systems. *Science*, *325*(5939), 419–422.
Pahl-Wostl, C., Lebel, L., Knieper, C., & Nikitina, E. (2012). From applying panaceas to mastering complexity: Toward adaptive water governance in river basins. *Environmental Science & Policy*, *23*, 24–34.
Perrone, D., & Hornberger, G. M. (2014). Water, food, and energy security: Scrambling for resources or solutions? *WIREs Water*, *1*(1), 49–68.

Perrone, D., Murphy, J., & Hornberger, G. M. (2011). Gaining perspective on the water-energy nexus at the community scale. *Environmental Science and Technology*, *45*(10), 4228–4234.

Richter, B. D., Mathews, R., Harrison, D. L., & Wigington, R. (2003). Ecologically sustainable water management: Managing river flows for ecological integrity. *Ecological Applications*, *13*, 206–224.

Rugel, K., Jackson, C. R., Romeis, J. J., Golladay, S. W., Hicks, D. W., & Dowd, J. F. (2012). Effects of irrigation withdrawals on streamflows in a karst environment: Lower Flint River Basin, Georgia, USA. *Hydrological Processes*, *26*(4), 523–534.

Ruhl, J. B. (2005). Water wars, eastern style: Divvying up the Apalachicola-Chattahoochee-Flint River Basin. *Journal of Contemporary Water Research and Education*, (131), 47–57.

Torak, L., & Painter, J. (2011). *Summary of the Georgia Agricultural Water Conservation and Metering Program and evaluation of methods used to collect and analyze irrigation data for the middle and lower Chattahoochee and Flint River basins, 2004–2010*. Reston, VA: US Geological Survey. https://pubs.usgs.gov/sir/2011/5126/pdf/sir2011-5126.pdf

University of Georgia. (2009). *2008 irrigation survey*. Tifton, Georgia 31793. http://caes2.caes.uga.edu/engineering/pubs/documents/irrigation survey08.pdf

US Bureau of Reclamation. (2012). *Klamath Tribes sociocultural/socioeconomics effects analysis technical report*. Denver, CO: US Department of the Interior. https://klamathrestoration.gov/sites/klamathrestoration.gov/files/2013%20Updates/Econ%20Studies%20/03.KlamathTribes_7AUG12.pdf

APPENDIX

Units, Dimensions, and Conversions

Hydrological Quantities

We can separate the quantities that we encounter in water resources into two classes. *Basic measurements* constitute the first class. For example, we might measure the height of water in a water well or the temperature of water in a lake. The second class includes *derived quantities*, because they are not measured directly but are calculated from measured variables using an equation representing a relationship between variables. For example, the discharge in a stream is calculated from the mean velocity and stream cross-sectional area.

Regardless of their type, hydrological quantities have an associated unit, such as centimeters per second, and a dimension, such as length per time (although some quantities may be unitless and dimensionless). In recording basic measurements or manipulating these values to derive other quantities, we need to be concerned with several things, such as *precision* and the appropriate units and dimensions, which are the subjects of this appendix.

Units and Dimensions

Consider a simple example. In hydrologic sciences we use the term "catchment" to indicate an area where water collects. A field investigation in a small catchment records some basic information, including the depth of water in a single precipitation collector placed within the catchment. We would like to know, as *precisely* and *accurately* as we can, with the available data, the total amount (volume) of water delivered to the catchment during a single event:

$$\text{Volume} = \text{Depth} \times \text{Area}$$

The measure of precipitation has an associated *dimension*, in this case length or [L], and *unit*, for example, centimeters or inches. We use as fundamental dimensions length [L], mass [M], and time [T] (Table A1). Many quantities have a dimension that is some combination of these fundamental dimensions. For example, the volume of precipitation falling on a catchment has dimensions of length cubed [L^3]. To continue with our calculation, we can first check to see if the relationship between depth and volume is *dimensionally homogeneous*—that is, whether the dimensions on both sides of the above equation are the same:

$$[L^3] = [L] \times [L^2] = [L^3]$$

As we would expect, this equation is dimensionally homogeneous. Dimensional homogeneity does not guarantee that the equation will give an accurate or correct result; for example, what if the length of the stream were used erroneously in place of the precipitation depth in the equation? However, an equation that completely and accurately describes a physical relation must be dimensionally homogeneous. It is always useful to check that any equation you use is dimensionally homogeneous.

There are a variety of units that correspond to each dimensional quantity. Length might be given in meters, centimeters, inches, or even gallons per square foot. Therefore, care must be taken to make sure any calculation is based on a *unitarily homogeneous* form of an equation. For the example, here is the precipitation over a catchment,

$$\text{Volume (m}^3) = \text{Depth (m)} \times \text{Area (m}^2)$$

Needless to say, a wrong answer will result if the units on one side of an equation are different from those on the other side.

The most common system of units employed today is the SI (*Système International d'Unités*; see Table A1). Other widely used systems of units include the *English system* (foot, pound, second) and the *cgs system* (centimeter, gram, second). The SI system includes a sequence of standard prefixes to indicate magnitude (Table A2). For example, a kilogram is equal to 1,000 grams, and a milligram is equal to 10^{-3} grams.

Table A1 Base and derived units relevant to water resources in SI measurement.

Quantity	Dimension	Unit	SI symbol
Base units			
Length	$[L]$	Meter	m
Mass	$[M]$	Kilogram	kg
Time	$[T]$	Second	s
Derived units			
Area	$[L^2]$	Square meter	m^2
Volume	$[L^3]$	Cubic meter	m^3
Discharge	$[L^3/T]$	Cubic meter per second	m^3/s

Table A2 Prefixes used in the SI system.

Prefix	SI symbol	Multiplication factor
Mega	M	$1{,}000{,}000 = 10^6$
Kilo	k	$1{,}000 = 10^3$
Hecto[a]	h	$100 = 10^2$
Deka[a]	da	$10 = 10^1$
Deci[a]	d	$0.1 = 10^{-1}$
Centi[a]	c	$0.01 = 10^{-2}$
Milli	m	$0.001 = 10^{-3}$

[a] Avoid use of this prefix where possible.

Significant Figures and Precision

To complete the simple example with which we began, we will use measured quantities to derive the volume of precipitation (in m^3) given a precipitation depth of 13 mm (0.013 m) and a catchment area of 2.065×10^5 m^2. The volume of water received by the catchment is 0.013 m $\times$ 2.065×10^5 $m^2 = 2{,}684.5$ m^3. This answer is correct mathematically, but it is expressed *with greater relative precision* than is justified by the measured values. There are too many significant figures in the answer, which requires some explanation.

First, the measured quantities have a certain *absolute precision*. For the measured precipitation depth, the absolute precision is 1 mm. It is likely that the rain gage could only measure precipitation depth *to the nearest millimeter*. The absolute precision for the area (206,500 m^2) is 100 m^2. This is apparent because we wrote the value initially using scientific notation. For a number written as 206,500, it is not clear whether the absolute precision is 100 m, 10 m, or 1 m. The *relative precision* is best thought of in terms of *significant figures*. In a given quantity, a significant figure is any

Table A3 Examples of quantities and their significant figures.

Quantity	Significant figures	Scientific notation
650,000	2[a]	6.5×10^5
30	1[a]	3×10^1
30.	2	3.0×10^1
30.0	3	3.00×10^1
30.01	4	3.001×10^1
.01	1	1×10^{-2}
0.01	1	1×10^{-2}
0.010	2	1.0×10^{-2}
0.00500	3	5.00×10^{-3}
1,000.0010	8	1.0000010×10^3

[a] Trailing zeros without a decimal point are ambiguous.

given digit, except for zeros to the left of the first nonzero digit, that serves only to fix the position of the decimal point. Some examples are given in Table A3.

There are a number of rules for dealing with derived quantities to ensure that the answers are expressed with the correct relative precision, or number of significant figures. When multiplying or dividing numbers, the answer should be expressed using the same number of significant figures as *the least relatively precise number involved in the calculation*, that is, the one with the fewest significant figures. In our example, the precipitation depth (13 mm) has two significant figures, and the catchment area (2.065×10^5 m^2) has four. Therefore, our answer should have only two significant figures: 2.7×10^3 m^3.

When adding or subtracting numbers, the rule is that the answer should be expressed using the same number of significant figures as *the number with the fewest decimal places*. Finally, numbers should not be rounded to the appropriate precision in a calculation with several steps *until the very end*.

Unit Conversions

Measurements that describe hydrological quantities may be expressed in a variety of different units. As a result, one often has to convert a quantity from one unit to another. For example, suppose that a value of discharge is given as 6.2 acre-feet per day. We would like to express this value in cubic meters per second. One approach is to multiply the value by ratios of equivalent units. The

Table A4 Equivalent units for length.

	Equivalent[a]					
Unit	millimeter	inch	foot	*meter*	kilometer	mile
millimeter	1	0.03937	0.003281	0.001000	1.0×10^{-6}	0.6214×10^{-6}
inch	25.4	1	0.0833	0.02540	25.4×10^{-6}	15.78×10^{-6}
foot	304.8	12	1	0.3048	304.6×10^{-6}	189.4×10^{-6}
meter	1,000	39.37	3.281	1	0.001	621.4×10^{-6}
kilometer	1,000,000	39,370	3,281	1,000	1	0.6214
mile	1,609,000	63,360	5,280	1,609	1.609	1

[a] In Tables A4 through A7, values are shown to four significant figures, and the SI expression, in base or derived units, is in italics.

Table A5 Equivalent units for area.

	Equivalent						
Unit	inch^2	foot^2	*meter*^{2}	acre	hectare	kilometer^2	mile^2
inch^2	1	0.006944	645.2×10^{-6}	15.94×10^{-8}	64.52×10^{-9}	645.2×10^{-12}	249.1×10^{-12}
foot^2	144	1	929.0×10^{-4}	22.96×10^{-6}	9.290×10^{-9}	92.90×10^{-9}	35.87×10^{-9}
meter^2	1,550	10.76	1	247.1×10^{-6}	10^{-4}	10^{-6}	386.1×10^{-9}
acre	6.273×10^{6}	4.356×10^{4}	4,047	1	0.4047	0.004047	0.001563
hectare	1.550×10^{7}	1.076×10^{5}	10^{4}	2.471	1	0.01	0.003861
kilometer^2	1.550×10^{9}	1.076×10^{7}	10^{6}	247.1	100	1	0.3861
mile^2	4.014×10^{9}	2.788×10^{7}	2.590×10^{6}	640	259	2.590	1

Table A6 Equivalent units for volume.

	Equivalent						
Unit	inch^3	liter	US gallon	foot^3	yard^3	*meter*^{3}	acre-foot
inch^3	1	0.01639	0.004329	578.7×10^{-6}	21.43×10^{-6}	16.39×10^{-6}	13.29×10^{-9}
liter	61.02	1	0.2642	0.03531	0.001308	10^{-3}	810.6×10^{-9}
US gallon	231.0	3.785	1	0.1337	0.004951	0.003785	3.068×10^{-6}
foot^3	1,728	28.32	7.481	1	0.03704	0.02832	22.96×10^{-6}
yard^3	46,660	764.6	202.0	27	1	0.7646	619.8×10^{-6}
meter^3	61,020	1,000	264.2	35.31	1.308	1	810.6×10^{-6}
acre-foot	75.27×10^{6}	1.233×10^{6}	3.259×10^{5}	4.356×10^{5}	1,613	1,233	1

Table A7 Equivalent units for discharge.

	Equivalent				
Unit	gallon minute^{-1}	liter second^{-1}	acre-foot day^{-1}	foot^{3}second^{-1}	*meter*^{3}*second*$^{-1}$
gallon minute^{-1}	1	0.06309	0.004419	0.002228	63.09×10^{-6}
liter second^{-1}	15.85	1	0.07005	0.03531	10^{-3}
acre-foot day^{-1}	226.3	14.28	1	0.5042	0.01428
foot^{3}second^{-1}	448.8	28.32	1.983	1	0.02832
meter^3*second*$^{-1}$	15,850	1000	70.04	35.31	1

ratios are formed such that the old units cancel, leaving the new units. The procedure is illustrated below:

$$\frac{6.2 \text{ acre–feet day}^{-1}}{1} \times \frac{1 \text{ m}^3\text{sec}^{-1}}{70.04 \text{ acre–feet day}^{-1}}$$
$$= 8.9 \times 10^{-2} \text{ m}^3\text{sec}^{-1}$$

Note that the answer is expressed in scientific notation, using the same number of significant figures (two) as the least relatively precise number, which in this case is the discharge value. It should also be obvious that the multiplication has resulted in the original units (acre-feet, day) being "canceled."

Tables A4 through A7 provide equivalent values for many quantities used frequently in water resources. To use the tables, look down the left column to find the unit you are interested in using to express a quantity. For example, if we wanted to convert acre-feet per day to cubic meters per second, we would look down the first column of Table A7 to find a row beginning with "meter^{3}second^{-1}." Scanning across this row, you will find that 1 meter^{3}second^{-1} is equivalent to 70.04 acre-feet day^{-1}, which is the value used to construct the conversion.

GLOSSARY

Albedo: the fraction of incident light that is reflected. (Section 10.2)

Anisohydric: descriptor of plants that tend to maintain the stomata open and sustain higher photosynthetic uptakes even when soil moisture is low. (Section 4.3)

Annual flood: the maximum recorded discharge at a stream gage in any given year. (Section 2.4)

Anthropogenic: originating from activity of humans. (Section 2.2)

Aquifer: a saturated geological formation that contains and transmits significant quantities of water under normal field conditions. (Section 1.5)

Assimilative capacity: the amount of waste that a receiving water body can disperse or degrade without unacceptable negative impacts. (Section 12.2)

Available moisture content: the moisture content between permanent wilting and field capacity that is available to plants. (Section 4.3)

Available water capacity (AWC): the amount of water that can be retained (not "lost" to deep percolation to the water table) by soils in the rooting zone of plants. (Section 1.5)

Average residence time: $T_r = V/I$, a measure of the average time a molecule of water spends in a reservoir. The residence time defined for steady-state systems is equal to the reservoir volume (V) divided by the inflow or outflow rate (I). (Section 1.7)

AWC: see **available water capacity**.

Baseflow: background low-flow conditions in a stream. (Section 1.3)

Biochemical oxygen demand (BOD): amount of oxygen per volume required to metabolize organic waste. (Section 12.2)

Biodiversity: a variety of species within a particular ecosystem or a variety of ecosystems within a region. (Section 9.3)

BOD: see **biochemical oxygen demand.**

Catchment area: an area of land, bounded by a divide, in which water flowing across the surface will drain into a stream or river and flow out of the area through a specified point on that stream or river. (Section 2.3)

Clausius-Clapeyron relationship: an equation that relates the water vapor pressure (a measure of how much water the atmosphere can hold) to the air temperature. (Section 10.3)

Cone of depression: the spatial pattern of drawdown in an aquifer as the result of pumping. (Section 3.3)

Confined aquifer: a permeable formation whose upper boundary is an impermeable formation. (Section 3.2)

Conservation(al) agriculture: type of agriculture focused on soil ecosystem health and sustainability; characterized by little to no tillage, cover crops, and crop diversification. (Section 5.1)

Consumptive use: the water that is removed from a source but is returned to the atmosphere by evaporation or transpiration rather than being returned to the surface as liquid water. (Section 2.3)

Conventional agriculture: type of agriculture requiring significant investments in mechanized energy or irrigation systems; sometimes referred to as industrial agriculture because of its high efficiency in labor and, consequently, large-scale farms. (Section 5.1)

Crop water need: depth (mm) of water needed to replace the water evapotranspired. (Section 5.6)

Deep percolation: the downward movement of water toward the water table below the plant rooting zone. (Section 1.3)

Discharge: the rate at which water flows expressed as a volume per time (e.g., m^3/s or cumecs). (Section 2.3)

Discharge hydrograph: a continuous record of river discharge as a function of time. (Section 2.3)

Drainable porosity: the fraction of pore water removed when the water table is lowered. (Section 3.3)

Drawdown: the change in water level in a well due to pumping. (Section 3.3)

Dry cooling: type of thermoelectric cooling that requires minimal water to be withdrawn or consumed compared to once-through or recirculating systems, because air is used to cool boiler water, not water. (Section 6.5)

Ecosystem services: the life-sustaining benefits that humans receive from natural ecosystem processes. (Section 2.2)

Eminent domain: government's right to take private property for public use by providing compensation. (Section 11.3)

Eutrophication: the over enrichment of natural waters with nutrients, which can lead to harmful algal blooms. (Section 12.2)

Evaporation: the physical process involving a phase change from liquid to vapor by which water is returned to the atmosphere. (Section 1.3)

Evapotranspiration: the sum of all processes, notably evaporation and transpiration, by which water changes phase (from solid or liquid) to vapor and is returned to the atmosphere. (Section 1.6)

Exceedance probability: the relative frequency associated with a random variable (e.g., the annual flood of a river) attaining a value greater than some specified value. (Section 2.4)

Field capacity: the relatively constant moisture content that a sandy soil tends to attain following drainage. (Section 4.3)

Flow duration curve: a graph that shows the fraction of time that river discharge is likely to equal or exceed some specified value. (Section 2.4)

Greenhouse gas: a gas that traps heat in the atmosphere by absorbing infrared radiant energy. (Section 10.1)

Groundwater: water found in the saturated zone of the subsurface. (Section 1.2)

Groundwater-dependent ecosystems: ecosystems that require groundwater flows for some or all of their survival requirements. (Section 8.2)

Groundwater depletion: the lowering of the water table or potentiometric surface of an aquifer due to pumping that exceeds the rate of recharge. (Section 3.2)

Groundwater mining: the removal of water from an aquifer that is not being recharged by the influx of water from outside the aquifer. (Section 3.4)

Hadley circulation: large scale global circulation in the atmosphere where warm air rises near the equator, flows away from the equator at a height of about 10 km, cools as it flows either northward or southward, descends to the surface at about 30 degrees N or S latitude, and flows back toward the equator at the surface. (Section 1.4)

Hydrograph: a continuous record of streamflow (stage or discharge) as a function of time. (Section 2.4)

Hydrological cycle: the global-scale, endless recirculatory process linking water in the atmosphere, on the continents, and in the oceans. (Section 1.1)

I = PAT: conceptually useful, but simplified, equation to evaluate the impact (I) of population (P), affluence (A), and technology (T). (Section 9.3)

Infiltration: the movement of rain or melting snow into the soil at Earth's surface. (Section 1.3)

Infrared radiation: radiation that is invisible to the eye but can be sensed as warmth by the skin. (Section 10.1)

Infrastructure: physical and organizational facilities required to provide services such as the delivery of potable water to communities. (Section 7.2)

Interannual variability: the year-to-year variation in a climate variable such as temperature, precipitation, or river discharge. (Section 10.3)

Interception: precipitation (either liquid or solid or both) temporarily stored on vegetation surfaces. (Section 1.3)

Intertropical convergence zone (ITCZ): a belt of low atmospheric pressure near the equator where surface winds from the northern and southern hemispheres flow together. (Section 1.4)

Irrigation efficiency: ratio of water conveyance efficiency and water application efficiency. (Section 5.2)

Isohydric: descriptor of plants that tend to close the stomata and reduce photosynthesis as the soil becomes drier. (Section 4.3)

ITCZ: see **intertropical convergence zone.**

Malthusian catastrophe: global population increases faster than global food production, ultimately leading to famine, starvation, and suffering. (Section 9.3)

Managed aquifer recharge: the intentional recharge of surface water into an aquifer to store it so it can be used at a later time. (Section 3.5)

Microirrigation: artificially applying small amounts of water over time to crop root zones; high irrigation efficiency. (Section 5.2)

Municipal water system: a system to provide water to municipalities that includes water storage facilities, water treatment facilities, and a water distribution network. (Section 7.2)

Nonpoint discharges: discharges that occur across a broad section of land through surface runoff or groundwater runoff. (Section 12.1)

Nonrenewable energy: energy sources that require special environmental conditions to form and take millions of years to develop; e.g., coal, natural gas, oil, and uranium for nuclear energy. (Section 6.1)

Once-through cooling: type of thermoelectric cooling technology that withdraws large amounts of water, runs the water through tubes in a condenser, and then discharges the water back to its source; withdraws significantly more water than recirculating cooling technology but consumes less water than recirculating cooling technology. (Section 6.5)

Organic agriculture: type of agriculture with well-defined certifications that emphasize the use of natural inputs or emphasis limiting synthetic inputs. (Section 5.1)

Orographic precipitation: precipitation resulting from upward motion of air caused by mountains; enhanced annual total precipitation typically is observed in many mountain ranges around the world. (Section 1.4)

Permanent wilting point: the driest soil moisture conditions that a plant can withstand without wilting. (Section 4.3)

PET: see **potential evapotranspiration.**

Point discharges: discharges that occur at one place through a pipe or a ditch. (Section 12.1)

Porosity: the fraction of the total volume of a rock or soil occupied by void space. (Section 3.2)

Potential: the hydraulic head due to gravity and pressure, the gradient of which drives the flow of groundwater. (Section 3.2)

Potential evapotranspiration (PET): the maximum rate of evapotranspiration from a vegetated surface under conditions of unlimited moisture supply. (Section 1.6)

Potentiometric surface: a surface that depicts the distribution of hydraulic heads in a confined aquifer. (Section 3.3)

Prior appropriation: water rights for the beneficial use of surface water or groundwater allocated by priority of the first use of the water; reduced allocations during times of scarcity are felt by junior rights holders first. (Section 11.3)

Radiant energy: energy that is transferred by electromagnetic radiation such as visible light, X-rays, and thermal radiation. (Sections 10.1 and 10.2)

Rating curve: a relationship between stage and discharge used to convert continuous measurements of stream depth (stage hydrograph) to a discharge hydrograph. (Section 2.3)

Recirculating cooling: type of thermoelectric cooling technology that withdraws water, runs it through tubes in a condenser, and then recirculates the water through the tubes in the condenser. Water withdrawals are low compared to once-through cooling technologies. To recirculate the water, heat is dissipated from the water to the atmosphere using cooling towers or ponds; this results in more water consumption compared to once-through cooling technologies. (Section 6.5)

Regulated riparianism: legislative approach to pure riparianism, requiring permitting, or permission, before water is withdrawn for some or all water users. (Section 11.3)

Regulations: rules set by a government agency to implement legislation that have the force of law. (Section 11.1)

Renewable energy: energy sources that can be replenished continuously within human timescales; e.g., solar, geothermal, wind, biomass, and hydropower. (Section 6.1)

Riparianism: water rights for the reasonable use of surface water allocated to water users adjacent to streams, rivers, or other water resources; users share the burden of reduced allocations during times of scarcity. (Section 11.3)

River stage: the depth of flow in a stream. (Section 2.3)

Safe yield: the rate at which water can be withdrawn from an aquifer without causing unacceptable harm to ecosystems or the environment. (Section 3.2)

Specific yield: the volume of water produced from an unconfined aquifer per unit aquifer area per unit decline in the water table. (Section 3.3)

Sprinkler irrigation: artificially applying water to crops via high-pressure sprinklers that spray water from the top-downwards. (Section 5.2)

Stage hydrograph: a continuous record of river stage as a function of time. (Section 2.3)

Statutes: laws enacted by the legislative branch of government. (Section 11.2)

Stomata: tiny pores in the leaves of vascular plants by which gases (including carbon dioxide, oxygen, and water) are exchanged with the atmosphere. (Section 4.3)

Storativity: the volume of water produced from a confined aquifer per unit aquifer area per unit decline in the potentiometric surface. (Section 3.3)

Surface irrigation: artificially applying water at the highest parts of a field so that if flows down across the field. (Section 5.4)

Surface runoff: water from rainfall or snowmelt that runs over the surface of Earth in sheets, rivulets, streams, and rivers. (Section 1.3)

Sustainable Development Goals: a set of goals established by the United Nations Development Program designed to eliminate poverty and protect the environment. (Section 7.1)

Takings: private property taken for public use (requires compensation). (Section 11.3)

Thermal pollution: negative impact of the discharge of water that is notably warmer than its natural conditions. (Section 12.1)

Thermoelectric: energy technologies that use heat to generate electricity by (1) burning an energy source (e.g., coal, natural gas, biomass) and converting the source to heat; (2) using fission to split uranium (nuclear) to release heat; or (3) using natural heat sources (e.g., heat from Earth's crust [geothermal] or heat from the sun [solar thermal]). (Section 6.5)

Tillage: process of breaking up and mixing topsoil to destroy weeds and pests and to redistribute and release nutrients from past harvest residues. (Section 5.1)

Trade-off frontier: a curve that depicts the production benefits associated with different mixes of water allocations when water is scarce enough to not be allocated to all demand-side sectors; for each gain in benefits in one sector, there must be a loss of benefits in another sector, that is, a trade-off. (Section 13.5)

Transpiration: the physical process by which water changes phase from liquid to vapor, is released through the stomata of a plant, and returns to the atmosphere. (Section 1.3)

Unconfined aquifer: a permeable formation whose upper boundary is the water table. (Section 3.2)

Unsaturated zone: the zone in a soil or rock between Earth's surface and the water table; pores in the unsaturated zone are partly filled with water and partly filled with air. (Section 4.1)

Virtual water: water embedded within food or other goods and services; this water can be water use throughout the good or services entire life cycle. (Section 9.4)

Volumetric moisture content: the volume of water held in a soil or rock per bulk volume of the sample. (Section 4.2)

Water conservation: a management technique that focuses on limiting the demand for water rather than increasing the supply of water. (Section 13.2)

Water governance: formal or informal rules under which management of water resources operates. (Section 13.2)

Waterlogging: saturation of the shallow soil resulting from the rising of the water table to the ground surface. (Section 2.4)

Water management: monitoring water uses and water levels, developing rules, and implementing measures to keep water use within identified boundaries. (Section 13.2)

Water scarcity: water supply does not meet water demand, which can be the cause of limited physical water supplies; limited investments in infrastructure or management policies; or political, cultural, or social constructs inefficiently influencing supplies and demands. Although traditionally tied to water quantity, water quality plays a role in the ability of supplies to meet demands. (Section 13.4)

Water table: a surface separating the saturated and unsaturated zones of the subsurface. (Section 3.2)

Water use efficiency: the ratio of the biomass produced by plants to the amount of water transpired. (Section 4.3)

Water year: the 12-month period from October 1 through the following September 30; the water year is expressed using the calendar year in which it ends. (Section 2.4)

Xylem: capillary tubes within a plant. (Section 4.3)

INDEX